Electronic Warfare for the Digitized Battlefield

For a listing of recent titles in the *Artech House Information Warfare Series,* turn to the back of this book.

Electronic Warfare for the Digitized Battlefield

Michael R. Frater
Michael Ryan

Artech House
Boston • London
www.artechhouse.com

Library of Congress Cataloging-in-Publication Data
Frater, Michael R.
Electronic warfare for the digitized battlefield / Michael R. Frater, Michael Ryan.
p. cm. — (Artech House information warfare library)
Includes bibliographical references and index.
ISBN 1-58053-271-3 (alk. paper)
1. Electronics in military engineering. 2. Command and control systems.
3. Information warfare. 4. Military art and science—Automation.
I. Ryan, M. J. (Michael J.) II. Title. III. Series.
UG485 .F73 2001
623'.043—dc21 2001026675

British Library Cataloguing in Publication Data
Frater, Michael R.
Electronic warfare for the digitized battlefield. — (Artech House information warfare library)
1. Electronics in military engineering 2. Communications, Military
I. Title II. Ryan, Michael
623'.043

ISBN 1-58053-271-3

Cover design by Gary Ragaglia

685 Canton Street
Norwood, MA 02062

International Standard Book Number: 1-58053-271-3
Library of Congress Catalog Card Number: 2001026675

10 9 8 7 6 5 4 3 2 1

Contents

Preface

Modern land commanders are increasingly dependent on Information Age systems comprising communications and information systems, networks, and sensors. While these systems have the potential to produce significant changes in the conduct and character of war, their reliance on the electromagnetic spectrum also has the potential to increase their vulnerability to interdiction by electronic warfare systems.

There have been many books and articles describing *noncommunications electronic warfare,* that is, electronic warfare in the context of electronic sensor systems, particularly radar. In this book we address the critical issues related to the effect of electronic warfare on the business of command and control on the digitized battlefield. More specifically, we address the effect of electronic warfare on the battlefield communications systems that support the command and control process. This aspect of electronic combat is called *communications electronic warfare.* Moreover, we focus on the components and techniques employed at the tactical level of land warfare, that is, at division and below. The engineering principles, technology, doctrine, and procedures contained in this book are drawn from open sources readily available in the public domain.

Chapter 1 describes the operational environment of the digitized battlefield and examines the process of command and control that is the core business of the tactical commander. The concept of network-centric warfare is discussed as an example of a doctrine that is emerging to harness the power of the information revolution for application to land warfare. This doctrine is then examined in the context of the heavy reliance that networked forms of warfare have on the use of the electromagnetic spectrum. The information revolution not only provides an improved ability to command and control, but also brings with it a commensurate ability to disrupt the process. The emerging doctrine of *information warfare, information operations,* and *command and control warfare* is then discussed to provide a framework within which to consider the role of electronic warfare on the digitized battlefield. A taxonomy is given for the doctrine of electronic warfare, comprising *electronic support, electronic attack,* and *electronic protection.*

Chapter 2 discusses the targets of tactical communications electronic warfare—the communications systems that underpin the ability of a tactical commander to command and control. Tactical trunk communications, combat net radio, and tactical data distribution systems are described, and an architecture is developed to illustrate the interrelationship of systems required to provide the battlefield networks that support operational concepts such as network-centric warfare.

Chapter 3 concentrates on electronic protection, which comprises those actions taken to protect friendly equipment from any effects of friendly or adversary use of the electromagnetic spectrum that degrades, neutralizes, or destroys friendly combat capability. This is achieved through a combination of active (detectable) and passive (undetectable) means.

In Chapter 4 we describe electronic support, which is the component of electronic warfare focused on the identification of sources of intentional and unintentional radiation of electromagnetic energy by an adversary. The elements of electronic support are *search, intercept, direction finding,* and *analysis.* Electronic support provides intelligence on adversary activity and deployment, as well as steerage for electronic attack.

Electronic attack is the focus of Chapter 5, which describes the issues associated with the use of electromagnetic energy to attack adversary equipment with the intent of degrading, neutralizing, or destroying adversary combat capability. Electronic attack can take the form of jamming, electronic deception, or neutralization.

Chapter 6 provides a description of the organization of EW units and the planning processes required for the deployment of electronic warfare capabilities on the battlefield. Effective planning for EW is crucial to overcome the scarcity of EW resources and to maintain security and the requirement to coordinate with other functional areas such as communications.

In Chapter 7 we discuss the emerging field of directed-energy weapons. While these weapons provide a means of electronic attack, they have been considered separately in this chapter due to the different nature of their attack and their potential to affect a much wider range of equipment through a broader range of attack mechanisms. The chapter also examines the means of protection, both tactical and technical, against the use of such weapons.

Finally, Chapter 8 addresses the future directions of battlefield electronic warfare systems as tactical communications continue to develop to take advantage of the opportunities offered by the information revolution. A key driver of future EW will be the way in which target tactical communications systems evolve towards a true battlefield network.

It has been several hundred years since commanders have had the ability from a convenient hilltop to survey with their own eyes the disposition of friendly and adversary forces. Now, in the Information Age, modern commanders, their senses enhanced by electronic sensors and modern communications systems, can stand on their electronic hilltop and once again "see" whatever portion of the battlefield is desired at whatever detail is appropriate. Yet, while the information revolution promises to change fundamentally the conduct and character of war, the increased reliance on electrical communications systems creates the potential for a greater vulnerability to electronic warfare. Greater investment is therefore required in offensive and defensive electronic warfare equipment, personnel, and training. As armies expend large sums of money seeking to attain the advantages of digitization, consideration must also be given to the flip side of the information revolution—the increased role of electronic warfare on the modern digitized battlefield.

1

The Digitized Battlefield

1.1 The Operational Environment

Throughout history, technological, political, and social advances have caused profound shifts in military doctrine, organization, strategy, and tactics. In recent history, six revolutions in military affairs have radically altered the conduct and character of war. The first five were the institution of universal military obligation (the French Revolution of 1789), the Industrial Revolution of the midnineteenth century, the managerial revolution of the late nineteenth century, the mechanized revolution occurring between 1919–1939, and the scientific revolution that followed shortly afterwards, culminating in the production of the atomic bomb. Then, in the early 1970s, the introduction of precision-strike weapons and computers produced the latest revolution—an information revolution centered on the concept that the dominant factor in war is the ability to collect, analyze, disseminate, and act upon battlefield information [1].

These advances in technology have produced an environment on the modern battlefield that is characterized by continuous, 24-hour action; increased volume, lethality, range, and precision of fire; smaller, more effective units due to better integration of technology; a disjunction between greater dispersion of more mobile, faster units and an increased tendency for combat in built-up areas with congestion of forces in short ranges; and a further dichotomy between greater invisibility, due to dispersion and speed, and increased risk of detection, due to larger numbers of more capable battlefield sensors.

Arguably, therefore, the most significant technological revolution in warfare will be in the role of information and knowledge, and, in particular, in the degree of situational awareness made possible by the increasing number of communications and information systems supporting combat forces. However, not all armies will be able (or will choose) to take advantage of this revolution, and today's Information Age army must be prepared to deal with a broad spectrum of threats from Agrarian, Industrial, and Information Age adversaries [2].

The information revolution, with the associated provision of information technology, favors networks rather than hierarchies; it diffuses and redistributes power; it crosses borders and redraws boundaries of offices and responsibilities; and it expands horizons. This is particularly true in the civilian environment, where organizations have become more democratic in information distribution and have realized considerable efficiencies [3].

> The network form is very different from the institutional form. While institutions (large ones in particular) are traditionally built around hierarchies and aim to act on their own, multi-organizational networks consist of (often small) organizations or parts of institutions that have linked together to act jointly. The information revolution favors the growth of such networks by making it possible for diverse, dispersed actors to communicate, consult, coordinate, and operate together across greater distances and on the basis of more and better information than ever before. [4]

For warfare, the major lesson from the commercial world is that in the Information Age conflict will largely be about knowledge, and mastery of the network and networked organizations will provide major advantage in conflict. However, these concepts can be an anathema to military commanders who tend to see command and information (and even communications in many armies) following the same hierarchical lines. In a nonhierarchical, network model, command and information flow must necessarily diverge. Sensors, commanders, and weapon systems are connected via a networked grid that ensures that situational awareness data can be shared by all elements, regardless of whether they belong to the same unit. Command lines need no longer be shared with information flow. Information is shared across the network; command and control are directed in accordance with the order of battle. Therefore, the adoption of these new technologies will not only significantly affect the way armies are commanded and controlled, but it will also change the way they are organized and trained.

There are many books and articles that address the issue of warfare in the Information Age [5]. However, we will focus on the framework articulated by the U.S. Joint Vision 2020 (JV2020) [6], which provides a useful background for our consideration of electronic warfare on the digitized battlefield. JV2020 builds upon the conceptual template established in Joint Vision 2010, and has the goal of transforming U.S. forces to create a power that is dominant across the full spectrum of military operations. Modern armed forces must be able to defeat adversaries across a wide range of operations such as conventional warfighting, peace enforcement, peacekeeping, counterterrorism, humanitarian assistance, and civil support. In this book, the term "battlefield" is used to refer generically to operations across the spectrum.

A key component of full-spectrum dominance is *information superiority*—the capability to collect, process, and disseminate an uninterrupted flow of information while exploiting or denying an adversary's ability to do the same. Information superiority can therefore be defined as "that degree of dominance in the information domain which permits the conduct of operations without effective opposition" [7]. Superior information is to be converted to superior knowledge which, when combined with organizational and doctrinal adaptation, relevant training and experience, and proper command and control mechanisms and tools, is to achieve *decision superiority.*

JV2020 proposes that current capabilities for maneuver, strike, logistics, and protection will become dominant maneuver, precision engagement, focused logistics, and full-dimensional protection. The following descriptions of these terms are taken from the definitions provided in JV2020.

- *Dominant maneuver.* This is defined as the ability of joint forces to gain positional advantage with decisive speed and overwhelming operational tempo in the achievement of assigned military tasks. Widely dispersed joint air, land, sea, amphibious, special operations, and space forces, capable of scaling and massing force or forces and the effects of fire as required for either combat or noncombat operations, will secure advantage across the range of military operations through the application of information, deception, engagement, mobility, and countermobility capabilities.
- *Precision engagement.* This is the ability of joint forces to locate, observe, discern, and track objectives or targets; select, organize, and use the correct systems; generate desired effects; assess results; and reengage with decisive speed and overwhelming operational tempo as required, throughout the full range of military operations.

- *Focused logistics.* This is the ability to provide the joint force the right personnel, equipment, and supplies in the right place, at the right time, and in the right quantity, across the full range of military operations. This will be made possible through real-time, networked information systems providing total asset visibility as part of a common relevant operational picture, effectively linking the operator and logistician across service and support agencies. Through transformational innovations to organizations and processes, focused logistics will provide the joint warfighter with support for all functions.
- *Full-dimensional protection.* This is the ability of the joint force to protect its personnel and other assets required to decisively execute assigned tasks. Full-dimensional protection is achieved through the tailored selection and application of multilayered active and passive measures, within the domains of air, land, sea, and space, and information across the range of military operations with an acceptable level of risk. The dimensions of protection range from forward-deployed forces, through supporting logistics, to home commands and supporting space surveillance and communications systems. One dimension of protection, for example, is the protection of forces at garrisons and military bases. Asymmetric terrorist attacks pose a threat that must be countered by layers of defense including active human intelligence (HUMINT) on terrorist activities, passive monitoring of traffic around the base, alert conditions and procedures for tightening perimeter security, covert intrusion detection sensors, facility decoys, and levels of physical security access.

JV2020 also places significant emphasis on *information operations* (those actions taken to affect an adversary's information and information systems while defending one's own information and information systems) as an essential element of achieving each of the elements of full-spectrum dominance. We will return to this topic shortly, as electronic warfare is an important component of information operations.

Recognition is also given in JV2020 to the fact that the adoption of information technologies is not sufficient to make maximum use of the opportunities made available by the information revolution. The vision of JV2020 is to be realized through a transformation of the necessary doctrine, organization, training, material, leadership and education, personnel, and facilities.

Perhaps the most useful elaboration of the impact of information technology is in the emerging concept of *network-centric warfare* (NCW).

In the current platform-centric warfare, the sensing and engaging capability normally resides on the weapon system ("shooter"), and there is only a limited capability for the weapon to engage targets because it can only use the situational awareness generated by its own sensor. If a weapon is able to engage a target located by a remote sensor, the passage of weapon data is normally via stovepipe communications systems (i.e., they connect the single weapon directly to the single sensor). By contrast, in network-centric warfare, sensors and shooters are connected by a ubiquitous network through which weapons can engage targets based on a situational awareness that is shared with other platforms. Combat power can therefore be applied with fewer weapons systems than are currently required. Note that, just because weapons and sensors are interconnected, it does not mean that targets can be engaged randomly or without authority; control is still essential to ensure that targets are engaged in accordance with the operational plan.

While there may continue to be a role for direct links from sensor to shooter, the ultimate aim of NCW is that the employment of future precision weapons is designed around information. No single sensor has the ability to direct the application of these precision weapons—data must be integrated from a number of sensors and databases. On the modern battlefield, the network is a considerable force multiplier. Commanders will be unfettered by communications and unconstrained to information centers (command posts). The information network must be ubiquitous across the battle space and must be fluid, flexible, robust, redundant, and real-time; have integrity and security; have access and capacity; and be joint- and coalition-capable.

In their discussion of the associated issues, Alberts, Gartska, and Stein define NCW as "an information superiority-enabled concept of operations that generates increased combat power by networking sensors, decision makers, and shooters to achieve shared situational awareness, increased speed of command, higher tempo of operations, greater lethality, increased survivability, and a degree of self-synchronization" [8].

Figure 1.1 illustrates the three interlocking grids of NCW (the *information grid,* the *sensor grid,* and the *engagement grid*), and the three major types of participants (*sensors, command elements,* and *shooters*). The information grid provides the infrastructure through which information is received, processed, transported, stored, and protected. The sensor grid contains all sensors, whether they are specialized devices mounted on weapons systems, carried by individual soldiers, or embedded into equipment. The engagement grid consists of all available weapons systems that are tasked to create the necessary battlefield effect. Proponents of NCW envisage that these three grids will exist in space, in the air, on land, and on and under the sea.

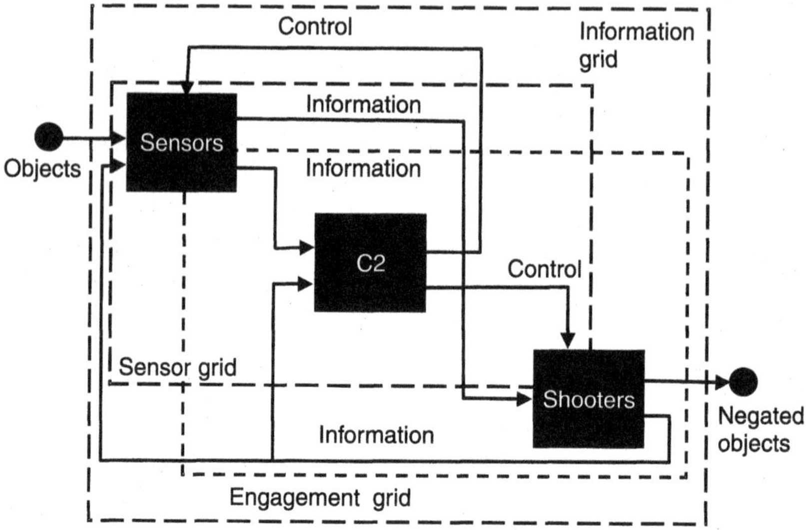

Figure 1.1 The grid arrangement of network-centric warfare [9].

NCW is not currently part of U.S. doctrine. However, the concept has considerable merit philosophically and it is most likely that future land warfare will embrace most, if not all, of the above concepts. The employment of a tactical network based on wireless, nonnodal communications has the advantage that armies can disperse as required and then mass effects rapidly at an appropriate time and place. Less reliance is required on large information processing centers, which can be distributed to increase physical survivability without sacrificing processing power.

This section has provided a very brief introduction to the operational environment of the future. While we have not considered in detail many of the issues associated with the significant impact that the information revolution will have on battlefield weapons systems, the most significant effect for our discussion of electronic warfare will be on the ability of a commander to acquire information, prepare and disseminate plans, and then control their execution. This is the business of *command and control,* which has become increasingly dependent on reliable communications and effective information systems. So, before we consider further the role of information warfare, particularly the role of electronic warfare, it is important to address the issue of command and control in more detail.

1.2 Command and Control

Command and control (and particularly its automation) is too broad a subject to be treated in detail here. Interested readers are referred to the many texts that cover the field [10]. For our purposes, we do not dwell on the detail but settle for the following working descriptions.

Command is perhaps best described as the authority vested in an individual for the direction, coordination, and control of military forces. Control is the means by which command is exercised. In a simple organization, the commander does most of the controlling; in a more complex organization, most of the control functions are delegated to supporting staff who form a headquarters supporting the commander. *Control* involves analysis of requirements, allocation of resources, integration of effort, direction, coordination, and monitoring.

The two terms command and control are inextricably linked—command is hollow without an ability to control; control is toothless without the authority of command. Therefore, the business of a commander is most often referred to as command and control (C2), which can be described as the process of and means for the exercise of authority of a commander over assigned forces in the accomplishment of the commander's mission. U.S. doctrine adds that command and control functions are performed through an arrangement of personnel, equipment, communications, facilities, and procedures employed by a commander in planning, directing, coordinating, and controlling forces and operations in the accomplishment of the mission [7].

1.2.1 The C2 Cycle

The interdependence of the various elements of a command system is illustrated by the C2 cycle shown in Figure 1.2. Although a very simple model, the C2 cycle is a useful mechanism for developing a framework for the application of command and control at any level. Here it is also useful to visualize the impact that communications and information systems have on the modern battlefield.

The C2 cycle is also called the *decision cycle,* the OODA (OUDA) loop (for the elements of *o*bservation, *o*rientation (*u*nderstanding), *d*ecision and *a*ction), or the *Boyd cycle* [11] (after the retired U.S. Air Force Colonel John Boyd who pioneered the concept).

While the cycle is continuous, it can be considered to start with surveillance and target acquisition (STA), or observation, from which the commander receives a wide range of information from the many sensors and systems deployed. This information is invariably reported in digital form, and the rapid increase in the number of sensors and surveillance systems is predominantly responsible for the explosion in digital transmission requirements on the modern battlefield. It should be noted that surveillance data can

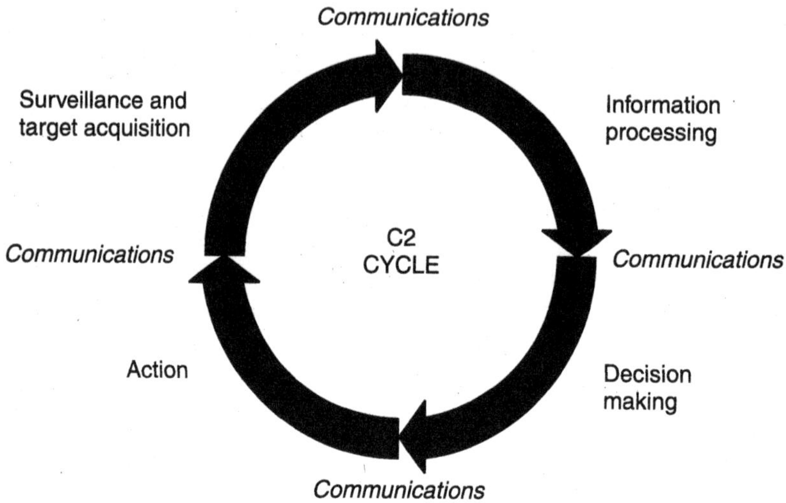

Figure 1.2 The C2 cycle.

only reach the commander if effective, survivable communications systems are available from the sensor system through to the data processing facilities in the command post.

The volume of raw sensor information coming into the headquarters is overwhelming and must be filtered and displayed in an appropriate format for the commander and staff to take action on it. As the volume of information grows, automation of this process is essential. It is estimated that a divisional commander currently has available something on the order of 1,000 times more information than would have been available in 1980. High-speed networks must be provided within a headquarters to facilitate data processing. The commander then makes a number of decisions and finalizes a plan, following which orders are conveyed to subordinate units through data and voice networks.

The purpose of all the preceding stages of the loop is to initiate action. The C2 cycle must assist the commander to take effective action based upon a correct appreciation. However, few plans last beyond first contact with the enemy so the loop must continue and information begins to flow to support the new operation. The commander must then control the action of subordinate units and task STA assets to monitor operations.

There are many more complex models for command and control. However, the C2 cycle is adequate for our purposes here, since it is evident even from the simple model that a major factor in success on the battlefield is the ability to move through the C2 cycle more quickly than an adversary.

It is here that the information revolution offers the greatest promise for improvement, albeit with a corresponding increase in vulnerability. We will return to that point shortly, but for the moment, we should note the heavy reliance that the C2 cycle has on technologies that require use of the electromagnetic spectrum.

The term *battlefield digitization* has been adopted to refer to the automation, through digital networks and processes, of command and control operations across the full breadth of the battle space. This integration of ground, air, and space nodes (sensors, communications, command, and weapons nodes) into a seamless digital network requires the fully compatible digital exchange of data and common operating pictures to all nodes. Security, compatibility, and interoperability factors dominate the drive toward full digitization across the entire battle space.

1.2.2 Command Systems

While the term C2 remains in common use to refer to the processes and means for the exercise of authority, the field has spawned many variations in terminology, for example: command, control, and communications (C3); communications and information systems (CIS); command, control, communications, and intelligence (C3I); command, control, communications, computers, and intelligence (C4I); or command, control, communications, computers, intelligence, surveillance, and reconnaissance (C4ISR). Each of these terms can be justified by its emphasis on particular vital elements of the command and control process. For example, without surveillance and reconnaissance, commanders are blind; without communications they are isolated, and so on. In the interest of brevity, we will simply consider the C2 cycle and bring together all of the systems that support it into the generic term of *command systems.*

To be successful on the modern battlefield, a commander and staff must be able to move through the C2 cycle faster than any adversary. Success in modern warfare depends on tempo, lethality, and survivability. Command systems must therefore be agile and responsive to changes in threat and must be able to cope with the influx of huge amounts of information from intelligence and surveillance systems, both tactical and strategic. In recent conflicts, this has overloaded tactical communications systems as well as the labor-intensive intelligence process, making it extremely difficult for the commander to process and analyze information in a timely manner. The implementation of automated battlefield information systems offers the only viable solution to process information and to prepare and disseminate plans within a realistic time frame.

A command system comprises automated and manual procedures to support a commander and staff. The essential components of a command system are the commander; supporting staffs; doctrine and procedures; reconnaissance and STA systems; communications systems; and information systems.

Arguably, the most important component is still the human element, comprising an able commander supported by well-drilled staff and appropriate doctrine and procedures. Despite the enormous advantages to be accrued from the information revolution, we must continue to be mindful that technology alone will not win battles, nor does the adoption of new technology obviate the need for the development of appropriate doctrine and procedures. History repeatedly warns us that it is not technology in itself that is important, but rather the way in which it is harnessed. Changes in doctrine, training, and organization have been necessary before any given technology has had any true battlefield effect [12]. As Van Crefeld observes: "[S]ince a decisive technological advantage is a fairly rare and always temporary phenomenon, victory often depends not so much on having superior technology at hand as on understanding the limits of any given technology, and on finding a way around these limitations" [13].

It must be also remembered that a command system comprises a set of *automated* and *manual* procedures. Sometimes manual procedures are more appropriate and consideration must be given to the implementation of a set of procedures that can survive the destruction or degradation of communications and information systems. While maintenance of a technological edge is vital, most command and control failures of this century have resulted from human mistakes rather than a lack of technology. Modern Western armies seem increasingly to imagine that precision maneuver enabled by information will create an overwhelming advantage. It must be remembered that the United States maintained an overwhelming technological and airpower advantage throughout the Vietnam War.

However, the implementation of information systems and information technology is essential to provide the automation necessary to transfer, process, and store the large volumes of data on the future battlefield. Technology will play a significant role in the support developed to allow commanders to plan and maneuver faster than an adversary. Information systems and technology might be expected to improve several thousand times in the next 20 years and will greatly increase the scope, volume, accuracy, and speed of information available to commanders.

1.3 Information Warfare

While the Information Age has produced a revolution in military operations that provides a great promise of decisive advantage on the modern battlefield to the commander who can gather and exploit information most effectively, there is a significant dark side to the information revolution. As communications and information systems become vital to military and civilian society, they can become major targets in war and can also serve as a major means for conducting offensive operations. Consequently, the military adoption of information technology creates a new vulnerability—the same information technology that provides the fuel for the networks that support modern commanders also provides one of the major means for their destruction. An increased reliance on communications and information systems increases this vulnerability. So, while automated command systems increase commanders' situational awareness, they can also be turned against them and used to contribute to their uncertainty.

It is evident from the preceding discussion that movement through the C2 cycle on the modern battlefield depends heavily on the use of the electromagnetic spectrum, whether for surveillance and target acquisition, passage of information, processing of information, or destruction of adversary forces. This reliance is a vulnerability that must be exploited in attacking adversary command systems, while being protected in own-force systems. Operations to counter the C2 cycle are generically termed *information warfare* (IW), which is a term that recognizes a range of actions taken during conflict to achieve information superiority over an adversary, and may be defined as [14]:

> Actions taken to achieve information superiority by affecting adversary information, information-based processes, information systems, and computer-based networks while defending one's own information, information-based processes, information systems, and computer-based networks.

The terminology and techniques of IW are ill-defined and without universal agreement, although there have been a number of comprehensive descriptions of the topic [15]. Again, we will not concern ourselves here with the detail of IW and related concepts, but limit our discussion to those aspects that impact on the field of electronic warfare.

The objective of IW is to attain a significant information advantage that enables the rapid domination and control of an adversary. The U.S.

Army recognizes that the current definition of IW is more narrowly focused on the impact of information during actual conflict and has chosen a somewhat broader approach to the impact of information on ground operations, adopting the term *information operations* (IO). IO integrate all aspects of information to support and enhance the elements of combat power, with the goal of dominating the battle space at the right time, at the right place, and with the right weapons or resources. Information operations are defined by FM100-6 as [16]:

> Continuous military operations within the military information environment that enable, enhance, and protect the friendly force's ability to collect, process, and act on information to achieve an advantage across the full range of military operations; IO include interacting with the global information environment and exploiting or denying an adversary's information and decision capabilities.

JV2020 adds that IO also include actions taken in a noncombat or ambiguous situation to protect one's own information and information systems as well as those taken to influence target information and information systems.

The warfighting application of IW in military operations is called *command and control warfare* (C2W). The aim of C2W is to influence, deny information to, degrade, or destroy adversary C2 capabilities while protecting C2 capabilities against such actions. C2W therefore comprises two major branches: C2-attack and C2-protect. C2W operations integrate and synchronize the capabilities of *psychological operations* (PSYOPS), *deception, operations security* (OPSEC), and *electronic warfare* (EW) [17].

It is the EW component, in particular, communications EW, that is of interest to us in this book. Although IW has the potential to have an impact much wider than the tactical environment, we focus on the warfighting application of communications EW on the digitized battlefield.

1.4 Electronic Warfare

Domination of the electromagnetic spectrum is a crucial component of most modern military operations. There are few battlefield elements that do not rely on communications and information systems. As discussed earlier, the C2 cycle depends very heavily on the electromagnetic spectrum to maximize the effectiveness of surveillance and target acquisition, communications, and information systems. If these systems are destroyed, degraded, or deceived, the commander and staff are unable to prosecute war adequately. For example, without communications on the modern battlefield, the commander is deaf, dumb, and blind. Therefore, the capability to conduct electronic combat and dominate the electromagnetic spectrum is now a recognized component of any modern force structure.

EW can be defined as "the use of the electromagnetic spectrum to degrade or destroy an adversary's combat capability (including degrading or preventing the use of the electromagnetic spectrum as well as degrading the performance of adversary equipment, personnel, and facilities); or to protect friendly combat capability (including protecting friendly use of the electromagnetic spectrum as well as friendly equipment, personnel, and facilities that may be vulnerable to attack via the electromagnetic spectrum)."

The targeting of personnel is beyond the scope of this book, which is focused on EW conducted against adversary communications and information systems. We therefore consider EW that is targeted against adversary communications, EW, and electronics. We also only consider EW as it is applied in the tactical context of the battlefield.

Figure 1.3 illustrates how electronic warfare pervades all aspects of the modern battlefield and has the potential to impact on all elements of the C2 cycle. In summary, EW resources are used to monitor adversary activities in the electromagnetic spectrum, indicate adversary strength and dispositions,

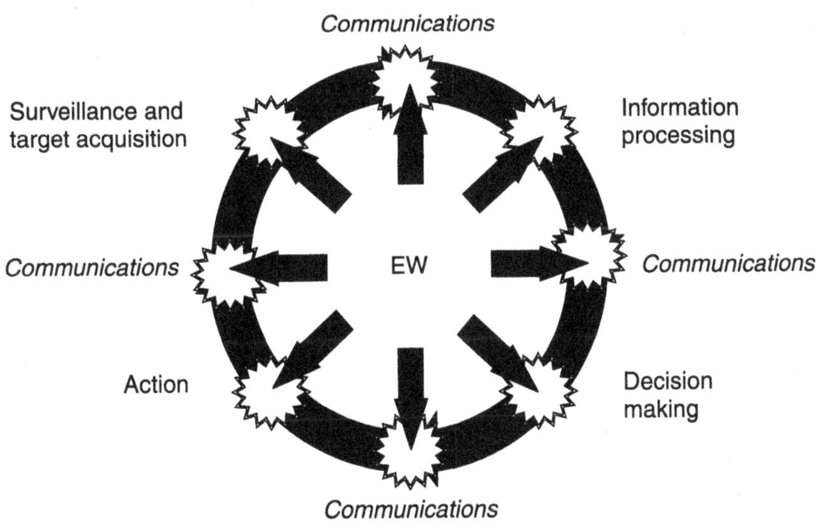

Figure 1.3 The potential impact of EW on the C2 cycle.

give warning of adversary intentions, deceive and disrupt sensors and command and control processes, and safeguard friendly use of the electromagnetic spectrum.

Although EW is targeted against the technology, the ultimate effect is on a commander's ability to move through the C2 cycle. The human element of the command system is both the strongest and weakest link and can fairly rapidly be enshrouded in the fog of war if supporting communications and information systems are disrupted, degraded, or deceived.

EW activities are applicable across the whole spectrum of military operations and are not confined to warfare, conventional or otherwise. In peacetime, armies attempt to intercept, locate, and identify the source of a potential adversary's electronic emissions. Analysis may then reveal details of capabilities as well as vulnerabilities that can be used to gain an advantage in times of conflict.

EW is an area of considerable innovation. Inevitably, and often very rapidly, advantages gained by technological or procedural change are met with equally effective countermeasures. In order to maintain the edge in any future conflict, information on friendly methods of electronic protection and attack must be safeguarded. Therefore, much of the parametric data associated with EW capabilities is highly classified. However, the underlying techniques and relationships can readily be obtained from open source publications.

1.4.1 Communications and Noncommunications EW

EW is normally divided into two main areas: *communications EW* and *noncommunications EW*. Communications EW is almost as old as electronic communications itself and, on the battlefield, is mostly concerned with communications sources that transmit in frequency bands between HF and SHF. The intercept and analysis of transmissions are usually more important than the measurement of transmitter characteristics. Noncommunications EW has been developed since the early employment of radars in World War II and is primarily concerned with platform protection, and normally specifically oriented towards radar systems in the UHF and higher bands. In noncommunications EW, the measurement of emitter characteristics is central as they are used to detect the presence of, and possibly identify, a piece of equipment and/or its performance.

As an aside, EW is also associated with *signals intelligence* (SIGINT), which contains two main subcomponents: *communications intelligence* (COMINT) and *electronic intelligence* (ELINT). To a large extent, these mirror the functional areas of communications and noncommunications EW, but take place in the strategic environment rather than the tactical one.

EW in the tactical land environment is mostly concerned with communications EW, which is therefore the focus of this book [18].

1.4.2 EW Subdivisions

As shown in Figure 1.4, there are three fundamental subdivisions within EW that are applicable to both communications and noncommunications EW, albeit with the different degrees of emphasis noted earlier:

- *Electronic support,* formerly known as *electronic support measures* (ESM), is the division of EW involving actions tasked by, or under the direct control of, an operational commander to search for, intercept, identify, and locate sources of intentional and unintentional radiated electromagnetic energy for the purposes of immediate threat recognition and constructing an *electronic order of battle* (EOB). An EOB includes information on the nature and the deployment of all electromagnetic emitting equipment of a military force, including equipment types, frequencies, modes of operation, locations, and other relevant data.
- *Electronic attack,* formerly known as *electronic countermeasures* (ECM), is the division of EW involving the use of electromagnetic energy to attack personnel, facilities, or equipment with the intent of degrading or destroying adversary combat capability. EA comprises *jamming, electronic deception,* and *neutralization.* Jamming is the use of electromagnetic energy to prevent a radio receiver from receiving its intended signal. Electronic deception involves the use of false or misleading transmissions to confuse an adversary. Neutralization describes the use of very high levels of electromagnetic energy to disrupt or permanently damage electronic equipment.

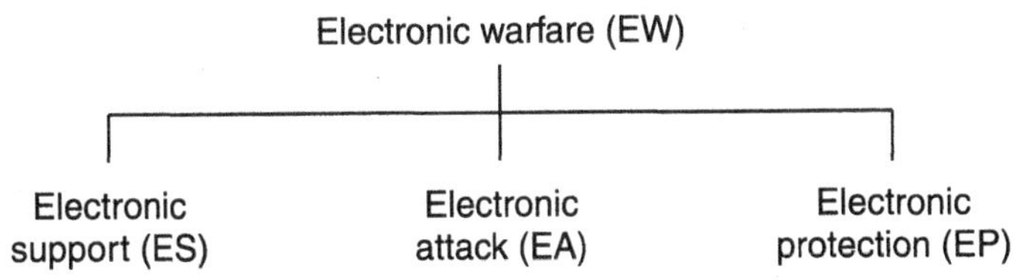

Figure 1.4 Major subdivisions of EW.

- *Electronic protection*, formerly known as *electronic protection measures* (EPM) or *electronic counter-countermeasures* (ECCM), comprises those actions taken to protect personnel, facilities, and equipment from any effects of friendly or adversary employment of EW that degrade, neutralize, or destroy friendly combat capability.

1.4.3 Other Categories of EW

EW can also be categorized as either *offensive* or *defensive.* ES and EA tend to be offensive, in that they are targeted toward an adversary and involve the process of *searching, intercepting, direction finding* (or *locating* or *position fixing*), *analyzing,* and engaging adversary electronic systems through jamming, deception, and neutralization. Mastery of offensive techniques, capabilities, and limitations is vital to the effective conduct of electronic combat. EP tends to be more defensive and protects own-force use of the electromagnetic spectrum against an adversary's offensive EW. EP is the concern of all users of electronic equipment and encompasses practices such as *emission security* (EMSEC) and *communications security* (COMSEC).

In turn, EW techniques can be characterized as either *passive* or *active* in nature. Passive activities are not detectable and can be implemented and practiced in peacetime with a limited risk of compromise. Active measures are detectable and should be carefully controlled on the battlefield and only permitted in peacetime under strict conditions. ES tends to be passive, while EA is active. EP contains both active and passive measures.

The diagram in Figure 1.5 gives an overall view of the interrelated activities associated with EW.

The remaining chapters of this book discuss each of these subdivisions in more detail, examining both the procedures and the characteristics of the equipment used.

1.5 Summary

Although the promise of command and control in the Information Age may stop short of completely dissipating the fog of war, it has the potential to turn night into day, to achieve spans of control that can be measured in global terms, and to mass collective combat power without massing forces [19]. The enduring lesson from recent conflicts since the Gulf War is that what can be seen can be hit, and what can be hit can be killed. The function of "seeing" is now much more sophisticated and entails electronic, optical,

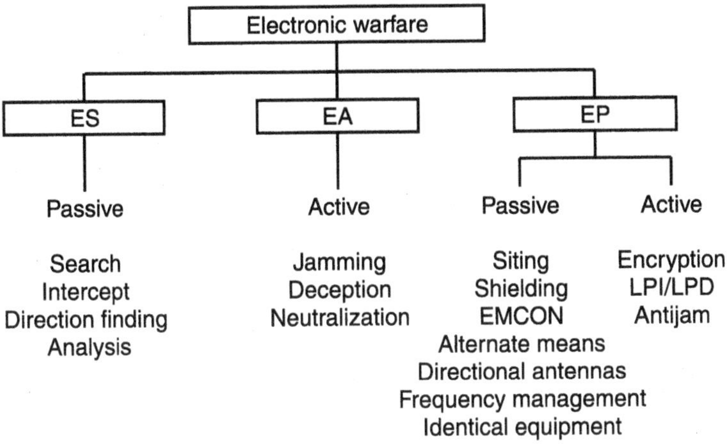

Figure 1.5 Overall view of EW.

and acoustic sensors that can have global coverage. These sensors can be linked in real time to computer-controlled weapon systems with unparalleled accuracy and lethality. However, this is not enough. The decisive advantage on the modern battlefield will go to the commander who can gather and exploit information most effectively. While this is greatly assisted by the technologies associated with the information revolution, the human element is arguably the most significant. For example, if computers and communications systems are used to reinforce hierarchical information flows—and therefore perpetuate the information overloads and bottlenecks—it is the fault of humans, not technology [20].

Commanders of the past have adopted most of their practices because the technology available simply did not allow them to do more. The information revolution will change that. Commanders can have unparalleled information available to them; they can "see" the full extent of the battlefield, even if it spans the globe. Careful thought must now be given to what practices are the most efficient. Just because it can be done, does not mean that it should.

Commanders will not have it all their own way, however. Future command and control systems will heavily rely on communications and information systems that cannot operate if access to the electromagnetic spectrum is denied. So, while the information revolution promises to deliver an enormous improvement in capability to commanders, it also creates the potential for new vulnerabilities. These new vulnerabilities offer new opportunities for the application of electronic warfare on the digitized battlefield.

One of the major impacts of the information revolution and increased use of networks is a significant broadening of the concept of battle space. As tactical commanders have increased access to information located anywhere in the tactical, operational, or strategic environments, they also become more vulnerable to IW assets located in any of those environments, not just the battlefield. This book is focused on the application of EW in the tactical environment and addresses those EW assets that have an intimate effect on the tactical commander.

Endnotes

[1] Reimer, D., "Foreword," in R. Pfaltzgraff and R. Shultz, (eds.), *War in the Information Age: New Challenges for U.S. Security Policy*, Washington, D.C.: Brassey's, 1997.

[2] Toffler, A., and H. Toffler, *War and Anti-War: Survival at the Dawn of the 21st Century*, Boston, MA: Little, Brown, 1993.

[3] Sproull, L., and S. Kiesler, *Connections: New Ways of Working in the Networked Organization*, Cambridge, MA: MIT Press, 1991.
Malone, T., and J. Rockart, "Computers, Networks, and the Corporation," *Scientific American*, September 1991, pp. 128–136.
Toffler, A., *Powershift: Knowledge, Wealth, and Violence at the Edge of the 21st Century*, New York: Bantam Books, 1990.

[4] Arquilla, J., and D. Ronfeldt, *Cyberwar Is Coming!* Santa Monica, CA: Rand, 1992, pp. 3–4.

[5] Further suggested reading on warfare in the Information Age:
Adams, J., *The Next World War*, London, U.K.: Random House, 1998.
Alexander, J., *Future War: Non-Lethal Weapons in Twenty-First Century Warfare*, New York: St. Martin's Press, 1999.
Allard, C., *Command, Control, and the Common Defense*, New Haven, CT: Yale University Press, 1990.
Arquilla, J., and D. Ronfeldt, (eds.), *In Athena's Camp: Preparing for Conflict in the Information Age*, Santa Monica, CA: Rand, 1997.
Campen, A., and D. Dearth, *CyberWar 2.0: Myths and Reality*, Fairfax, VA: AFCEA International Press, 1998.
Bellamy, C., *The Future of Land Warfare*, New York: St. Martin's Press, 1987.
De Landa, M., *War in the Age of Intelligent Machines*, New York: Zone Books, 1991.
Gordon, A., *The Rules of the Game: Jutland and British Naval Command*, Annapolis, MD: Naval Institute Press, 1996.
The International Institute for Strategic Studies (IIS), *Strategic Survey 1995–1996*, London, U.K.: Oxford University Press, 1996, p. 30.
Pfaltzgraff, R., and R. Shultz, (eds.), *War in the Information Age: New Challenges for U.S. Security Policy*, Washington, D.C.: Brassey's, 1997.
Rooney, D., V. Kallmeier, and G. Stevens, *Mission Command and Battlefield Digitization: Human Sciences Considerations*, DERA Report

DERA/CHS/HS3/CR980097/1.0, March 1998.
Scales, R. H., Jr., *Future Warfare: Anthology*, Carlisle Barracks, PA: U.S. Army War College, 1999.
Toffler, A., and H. Toffler, *War and Anti-War: Survival at the Dawn of the 21st Century*, Boston, MA: Little, Brown, 1993.
Van Trees, H., "C3 Systems Research: A Decade of Progress," in S.E. Johnson and A.H. Levis, (eds.), *Science of Command and Control: Coping with Complexity*, Fairfax, VA: AFCEA International Press, 1989.

[6] "Joint Vision 2020," Director of Strategic Plans and Policy, J5 Strategic Division, Washington, D.C.: U.S. Government Printing Office, June 2000.

[7] Joint Publication 1-02, "Department of Defense Dictionary of Military and Associated Terms," Washington, D.C.: Office of the Joint Chiefs of Staff, 1994 (as amended September 2000).

[8] Alberts, D., J. Gartska, and F. Stein, *Network Centric Warfare*, CCRP Publication Series, Washington, D.C.: U.S. Department of Defense, 1999.

[9] Cebrowski, A., and J. Garstka, "Network-Centric Warfare: Its Origins and Future," *Naval Institute Proceedings*, 1997.

[10] Further reading on command and control can be found in:
Dupuy, T., *Understanding War: History and Theory of Combat*, New York: Paragon House Publishers, 1987.
Echevarra, A., "Tomorrow's Army: The Challenge of Nonlinear Change," *Parameters*, Autumn 1998, p. 11.
Van Creveld, M., *Command in War*, Cambridge, MA: Harvard University Press, 1985.
Van Creveld, M., *The Transformation of War*, New York: Free Press, 1991.

[11] For a good description of the Boyd cycle see: Lind, W., *The Maneuver Warfare Handbook*, London, U.K.: Westview, 1985, p. 5.

[12] Builder, C., "Are We Looking in the Wrong Places?" in K. Thomas, (ed.), *The Revolution in Military Affairs: Warfare in the Information Age*, Canberra, Australia: Australian Defence Studies Centre, 1997, p. 7.

[13] Van Creveld, M., *Command in War*, Cambridge MA: Harvard University Press, 1985, p. 231.

[14] U.S. Army Field Manual FM100-6, "Information Operations," August 1996.

[15] Further reading can be found in:
U.S. Army Field Manual FM100-6, "Information Operations," August 1996.
Joint Publication 3-13, "Joint Doctrine for Information Operations," Washington, D.C.: Office of the Joint Chiefs of Staff, 1998.
Joint Publication 3-13.1, "Joint Doctrine for Command and Control Warfare (C2W)," Washington, D.C.: Office of the Joint Chiefs of Staff, 1996.
Libicki, M., *What Is Information Warfare?* Washington, D.C.: Institute for National Strategic Studies, National Defense University, 1996.
Rothrock, J., "Information Warfare: Time for Some Constructive Skepticism?" in J. Arquilla and D. Ronfeldt, (eds.), *In Athena's Camp: Preparing for Conflict in the Information Age*, Santa Monica, CA: Rand, 1997.

[16] U.S. Army Field Manual FM100-6, "Information Operations," August 1996.

[17] U.S. Army Field Manual FM100-6, "Information Operations," August 1996.

[18] Readers with interest in noncommunications EW are referred to:
Neri, F., *Introduction to Electronic Defense Systems*, Norwood, MA: Artech House, 1991.
Schleher, D., *Electronic Warfare in the Information Age*, Norwood, MA: Artech House, 1999.

[19] Allard, C., *Command, Control, and the Common Defense*, New Haven, CT: Yale University Press, 1990, pp. 263–264.

2

Tactical Communications Systems

2.1 Introduction

As discussed in Chapter 1, the pace and intensity of modern warfare require that commanders and staffs are supported by flexible, mobile, reliable, tactical communications systems that provide sufficient capacity to cope with increasingly high traffic loads in a hostile electromagnetic environment. Tactical communications systems have a similar form in all modern armies. To understand how tactical communications systems can be attacked by EW systems, it is important to understand their current structure, as well as the directions of their future development [1]. This chapter develops an architecture to illustrate the interrelationship of systems required to provide the battlefield network that supports operational concepts such as network-centric warfare. Current tactical trunk communications, combat net radio, tactical data distribution, and tactical airborne systems are then described in some detail. In Chapter 8 we address the future development of tactical communications systems.

2.2 Architectural Drivers

Over time, a number of principles for the provision of tactical communications have coalesced for communications support to land operations. These principles are well known and will not be dwelt upon here. Briefly, the major principles are that communications support the chain of command, integration, reliability, simplicity, capacity, quality, flexibility, anticipation of requirements, mobility, security, economy, survivability, and interoperability. In addition, for the modern battlefield, the following issues represent key design drivers for the provision of the tactical communications system.

Size of Supported Force

In the development of this architecture, attention is focused at the divisional level, that is, on division and below. While the architecture of the tactical communications system is developed with cognizance of larger formations, the most difficult problem is the provision of a sufficient digital capacity to tactical (division-level) combat forces, particularly at brigade and below.

Communications Support for the Spectrum of Operations

The communications architecture must be able to support a force in a range of operational deployments: from conventional, high-density motorized or mechanized operations to low-density operations, or in peacekeeping or peace-enforcing operations. To optimize procurement, training, and support, similar communications interfaces must be provided regardless of the type of operation. For small armies in particular, this will greatly ease procurement, logistic support, and training.

Command and Control on the Move

On the future battlefield it is essential that commanders can command and control on the move; a commander should be able to control force elements regardless of location. This includes continuity of command and control while deploying from barracks as well as while moving around the battle space. This requirement provides one of the most significant design drivers because it impacts on the power available for transmitters, the antenna size that can be supported, and so on.

Communications Support Situational Awareness

Communications must support situational awareness to commanders at all levels (in real time or near-real time) to provide accurate knowledge of adversary, friendly, neutral, and noncombatant entities. Modern warfare calls for near-real-time situational awareness at tactical headquarters, which provides a significant design driver for the tactical communications system. Traditionally, tactical communications have been based on the limited exchange of predominantly voice messages. The exchange of situational awareness information is based on data messages that must be shared across the whole battle space.

Seamless Connectivity

If concepts such as network-centric warfare are to be implemented on the future battlefield, the communications architecture must support seamless connectivity between any two points in the battle space, and between any point in the battle space and any point in the strategic communications system. While it will not always be desirable to be able to use such extremes of connectivity, the possibility of such connections must be supported in the architecture if the full power of network-centric warfare is to be realized.

Organic, Minimum-Essential Communications

The deployed force must have an organic, field-deployable tactical communications system that meets essential requirements for communications to support command and control. This minimum-essential system must provide guaranteed, robust, flexible communications that support the force, whether deployed on foot, motorized, or mechanized/armored. The tactical communications system is an organic asset that is part of the tactical force's combat power and can be guaranteed in any deployment. This also requires that the network is modular so that units and subunits are self-supporting when deployed separately from the main force and still retain communications functionality.

Expandable Communications

Since its organic communications will invariably be limited, the tactical force must be able to make use of other battle space and strategic communications systems when available. While essential force requirements are supplied by organic communications, additional capacity, redundancy, and reliability can be provided by using overlaid communications systems such as the commercial telephone network, satellite communications, theater broadcast, and so on. These systems must be seamlessly integrated with the tactical communications system.

Scalable Communications

Within all available assets, the tactical force must have the ability to provide scalable communications. That is, a small advance party must be able to deploy, taking with it sufficient communications for its task. As the force builds, the communications system must be able to grow to accommodate the size of the force.

Range

The tactical communications system must be able to support the force when it is deployed in any one of its roles. In the extreme, the communications system must be able to provide communications in conventional high-density deployments as well as support widely dispersed deployments. The tactical communications system must also be able to support high-capacity communications from the area of operations back into the strategic environment, which may require around-the-world communications.

Quality of Service (QoS)

The tactical communications system is required to carry a wide variety of traffic including voice, telegraph, video, and sensor data, reports, returns, navigation, location reporting, targeting data, database transfers, weather data, and so on. Increasingly, in modern digital networks, all of these forms of traffic are transmitted as data, albeit with varying requirements for speed of delivery and error rates. While warfare remains a human endeavor, commanders will want to communicate using voice. However, concepts such as network-centric warfare cannot be force multipliers unless data can be transferred rapidly across the battle space. Since there are many different types of data on the modern battlefield, it does not necessarily make sense to distinguish them by their source. Rather, it is more useful to characterize them by the requirements that each type of data has for services across the network. In that regard there are two main types of traffic: those that require real-time services (predominantly video and voice) and those that require non-real-time services (computer-to-computer transfer). Non-real-time services tend to be less tolerant of errors introduced into data during transmission.

Low Probability of Interception (LPI)

LPI, or *low probability of detection* (LPD), is critical to tactical communications systems. Survival on the modern battlefield requires the protection of communications systems as the first step in protecting the command systems that they support. LPI techniques include short-duration transmission; spread spectrum (direct sequence spread spectrum as well as frequency hopping); directional antennas; low-power settings; terrain screening; and use of airborne relays so that ground terminals can direct their power upward, away from a land-based enemy.

Jamming Resistance

Jamming resistance is also critical to the protection of communications links. Operation in a harsh electromagnetic environment requires the ability to implement measures to provide resistance to jamming. The techniques listed for LPI are also relevant to increase jamming resistance. Other techniques that may conflict with LPI requirements include increased power, strong error coding, jamming-resistant modulation, and adaptive antennas with steerable nulls.

These drivers are used as a framework in the following sections for the consideration of the development of architectures' tactical communications systems. Some issues such as LPI and jamming resistance are covered in more detail in Chapter 3.

2.3 Current Tactical Communications Architecture

As illustrated in Figure 2.1, in almost all modern armies the tactical communications system has evolved to comprise two major components: the *trunk communications* subsystem and the *combat net radio* (CNR) subsystem. [In U.S. doctrine, a third subsystem, the advanced data distribution system (ADDS), is provided—the utility of this third element is discussed in Section 2.7.] The trunk communications system provides high capacity links (terrestrial radio relay, satellite, fiber optic, or line) that interconnect headquarters at brigade level and above. The network is provided by a number of trunk

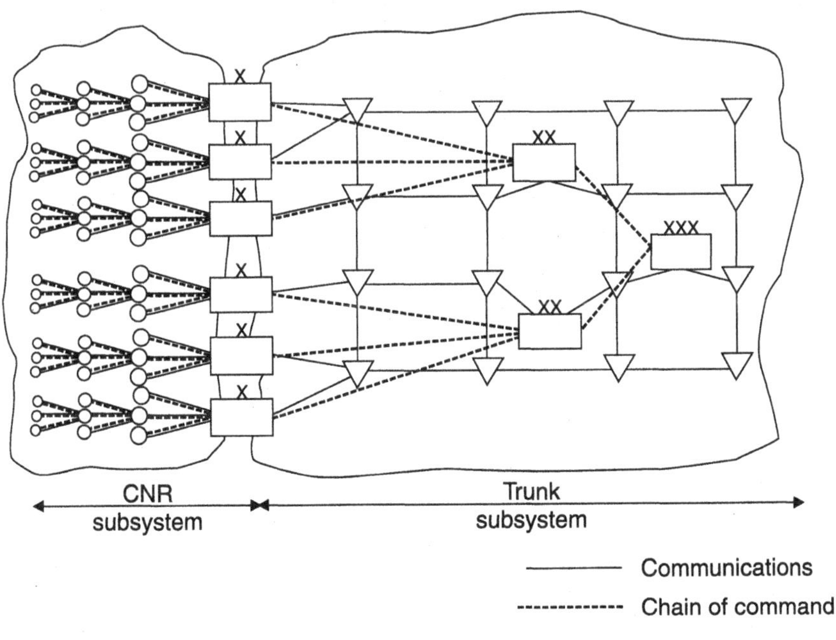

Figure 2.1 The architecture of the current tactical communications system.

nodes interconnected by trunk links, forming a meshed area network. Access is normally gained through access nodes that connect to one or more trunk nodes. Voice, telegraph, data, facsimile, and video facilities are provided to staff officers and commanders. The CNR subsystem is a ruggedized, portable radio network carried as an organic communications system for combat troops (brigade level and below). Radios are invariably interconnected to form single-frequency, half-duplex, all-informed, hierarchical nets, providing commanders with effective support to command and control.

There are a number of major problems with the current tactical communications system if it is to support command and control in future land warfare. The CNR subsystem is poorly placed to provide a network to transfer data between any two points in the battle space, due to its hierarchical net structure that has traditionally been devoted to analog voice traffic. The CNR and trunk communications subsystems are not seamlessly integrated to allow the transfer of information between any two points in the battle space. The tactical communications system is not seamlessly integrated with the strategic communications system to allow the transfer of information between any point in the battle space and any point in the strategic communications system. The data-handling capacity of the trunk communications subsystem will be sufficient (with some modification to the architecture) to cope with the volumes of data required to be transmitted between major headquarters. However, there is not a sufficient capacity below brigade (where trunk communications currently stop in most armies) to cope with future levels of data traffic required to provide commanders with near-real-time situational awareness. Finally, neither the CNR nor the trunk communications subsystems provide enough range to allow the dispersal of brigade elements when required. These issues and the key factors discussed earlier must be taken into account in the development of an architectural framework for a future tactical communications system.

The support for command and control in future land warfare requires the tactical communications system to be a single logical network to provide connectivity between any two points on the battlefield. The tactical communications system is an organic asset that provides the minimum-essential voice and data communications requirements to support situational awareness within the brigade and to allow for the transfer of command and control information. The tactical communications system interfaces with the strategic communications system to provide a seamless connectivity between any two points in the battle space and between any point in the battle space and any point in the strategic communications system.

2.4 A Suitable Tactical Communications Architecture for Future Land Warfare

The ideal tactical communications system architecture would provide a mobile infrastructure to support mobile users, a single homogeneous network supporting all communicating entities in the battle space. The tactical communications system is an organic asset that provides the minimum-essential voice and data communications requirements to support situational awareness within the brigade and to allow for the transfer of command and control information. The tactical communications system interfaces with the strategic communications system to provide seamless connectivity between any two points in the battle space and between any point in the battle space and any point in the strategic communications system.

The development of a suitable architecture for the tactical communications system can draw on the considerable body of knowledge available in existing commercial and military networks that provide mobile communications. However, in some respects, "mobile communications" is a misnomer when used in the commercial environment because only the user is mobile in such systems; the communications system (the network infrastructure) is very much fixed, with mobile access to this fixed infrastructure being provided by a wireless connection. In the military environment, the provision of a mobile communications system normally implies that both the user and the network infrastructure are mobile. Mobility therefore means markedly different things in the commercial and military environments. Consequently, while many commercial communications technologies are useful in the military environment, the mobility of the network infrastructure for military communications systems tends to require unique solutions.

While the tactical communications system can be provided as one logical network, it cannot be provided as one single physical network. At the lower level, combat troops carry a device that must be a network node as well as an access terminal. Battery power and the need for small, omnidirectional antennas mean that ranges and capacities are limited. At the higher level, the large capacities required of trunk communications systems mean that they will remain semimobile for the foreseeable future. Large power requirements must be met by the use of generators, and high-gain antennas must be deployed on guyed masts to provide reasonable ranges.

The data-handling capacity of the trunk communications system will generally be sufficient to cope with the volumes of data that must be transmitted between command posts. However, the CNR subsystem's ability is severely limited, especially as it is still required to transmit voice information. Therefore, an additional, purpose-designed, data distribution system is required to provide sufficient capacity to transfer situational awareness data across the lower levels of the battlefield. However, CNR must still be voice- and data-capable to allow organic communications of both types within subunits, should they be deployed individually or beyond the range of the data distribution subsystem. The additional (albeit limited) data capacity in the combat net radio subsystem would also provide an overflow capability should the tactical data distribution subsystem be unable to meet all the data needs.

Neither the CNR subsystem nor the trunk communications subsystem is able to cover the large ranges required for dispersed operation. The only solution to providing high-capacity, long-range communications is to elevate the antennas. In the extreme, the provision of a satellite-based or airborne repeater or switch will greatly increase the ranges between network nodes. A satellite-based solution is not considered desirable due to its inability to meet the requirements of a minimum organic communications system in most armies (even in those large armies that could afford integral satellite communications, such assets are likely to be provided sparingly and are relatively easily interdicted). An airborne subsystem is therefore required to support long-range operations. In addition, an airborne system will increase the capacity of lower-level tactical communications by removing the range restriction on high frequencies that can provide additional capacity from small, omnidirectional antennas.

By its very nature, a minimum organic tactical communications system will only be able to provide a basic level of service and must be able to be augmented where possible by overlaid communications systems such as the public telephone network, satellite-based systems, and personal communication systems. These overlaid systems cannot be guaranteed to be available and therefore cannot be included in the minimum organic system. If they are available, however, a great advantage is to be gained from their use.

In order to simplify the user interface to these subsystems, a local communications subsystem (most probably containing a level of switching) is required. This local subsystem could take a number of forms, from a vehicle harness to a local area network around brigade headquarters.

To support command and control in future land warfare, the tactical communications system is therefore required to evolve from the two subsystems in Figure 2.1 to five subsystems [2]. The *combat net radio subsystem* provides mobile infrastructure to carry voice and data communications to support the command and control of combat troops. The *tactical data distribution subsystem* provides high-capacity data communications to support the situational awareness required for the command and control of combat troops. The *tactical trunk subsystem* provides transportable infrastructure to support communications between command elements and other large-volume users. The *tactical airborne subsystem* extends communications ranges and provides additional capacity when the tactical situation allows. Finally, the *local subsystem* simplifies the user interface to the other communications subsystems and to overlaid communications systems.

The architecture of Figure 2.2 illustrates the major architectural components of the tactical communications system. It recognizes that, while the tactical communications system is to be considered as one logical network, for practical deployment reasons, it will be provided as a number of physical networks (at least in the short term). It is also a convenient starting point since it broadly coincides with the current deployed architecture, requiring

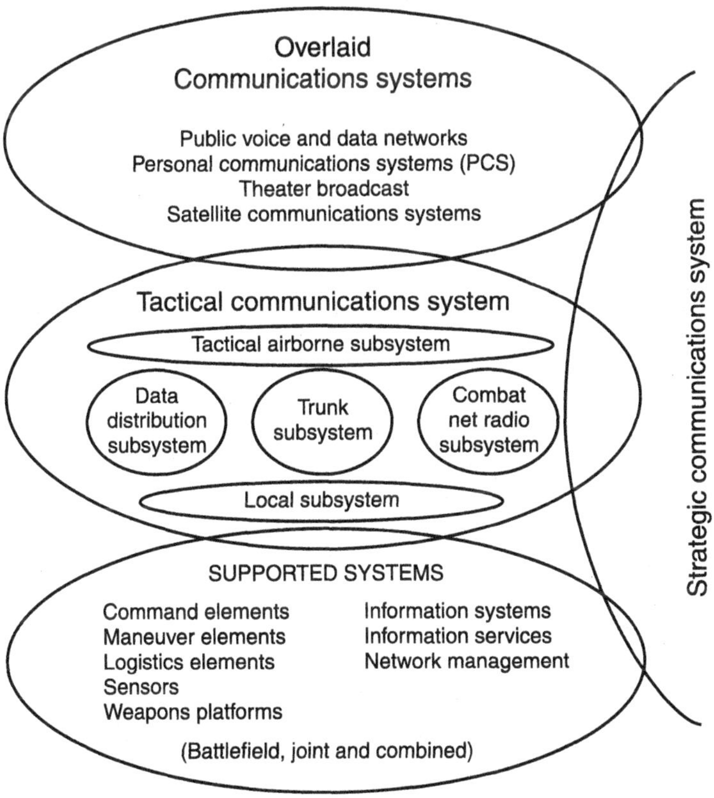

Figure 2.2 An architectural framework for the tactical communications system.

the addition of a tactical data distribution subsystem and a tactical airborne subsystem to increase capacity and range. However, the concept of a single logical network must remain paramount, as it is a crucial aspect of the architecture in Figure 2.2.

The tactical communications system does not exist in isolation; it exists to support a number of battlefield, joint, and combined systems. These supported systems interface to the tactical communications system as illustrated in Figure 2.2. As also illustrated in Figure 2.2, the minimum-essential tactical communications system is augmented where possible by a range of overlaid communications systems such as the commercial telephone network, satellite communications, theater broadcast, and personal communications system. These systems should be seamlessly integrated with the tactical communications system.

Here we are interested in those elements of the architecture that are principal targets for tactical communications EW, namely, the tactical trunk subsystem, the combat net radio subsystem, the tactical airborne subsystem, and the tactical data distribution subsystem. The following sections examine these subsystems in more detail.

2.5 Trunk Communications Subsystem

Trunk communications exist to interconnect command posts to allow the passage of large quantities of information between commanders and staffs. The three main forms of interconnections are *radio* (including radio-relay, tropospheric, and satellite bearers); *line* (including wire and fiber optic); and *hand carriage* (signal dispatch service). The main forms of traffic carried by trunk networks include voice, data, facsimile, telegraph, and data. The ability to use voice, their natural means of communication, is essential for commanders and staff. They also require the ability to transmit messages and facsimile for text transfer. Additionally, as the number of information systems increases, trunk networks must be able to transmit large volumes of data. The requirement to handle each of these forms of traffic efficiently is one of the major architectural drivers for tactical networks.

In addition to electronic forms of communication, most trunk networks include a hand-carriage component, as there are many communications that are too bulky and are best transferred between command posts by the signal dispatch service (SDS). Despite considerable advances, it is not likely that the technology of the information revolution will be able to provide the bandwidth to transmit all communications electronically.

Within a headquarters, commanders and staff officers are interconnected by their local subsystem to a local telephone exchange to provide connections to local subscribers as well as connections to the trunk switches through common-user trunk circuits. Data subscribers within a headquarters are interconnected by high-speed local area networks that are connected to each other across the trunk network.

Trunk communications systems provided the first forms of communication on the battlefield. Early systems relied on messengers to carry orders between a commander and subordinates. However, military communications requirements began to expand as fronts became wider, weapons became more sophisticated, and logistic tails became longer. To attack successfully, deploy reinforcements, commit reserves, and ward off counterattacks, commanders had to be aware instantly of events on distant battlefields—more swiftly than was possible by courier. Various visual and acoustic signaling systems were then deployed, including smoke, fire signals, heliograph, signaling lamp, and colored signaling flags. While these extended ranges and decreased the time taken to transmit simple messages, they were quickly replaced by the electrical telegraph, where the on-off keying of a simple electrical signal allowed the transmission of complex messages over long distances.

The first military use of the telegraph was during the Crimean War (1853–1856), and by World War I, it provided the mainstay of communications between headquarters. Tens of thousands of miles of telegraph cables had been laid by the end of the war. Radio telegraphy had been developed by the end of World War I, but it had been developed too late to be of any great use. By the beginning of World War II, however, radio had begun to provide high-capacity interconnections between headquarters, providing commanders with significantly greater freedom of movement, as they were no longer constrained by the cables connecting them with superior and subordinate commanders. By the end of World War II, radio had taken over as the principal bearer of trunk communications due to the greater mobility it offered commanders, with line being laid when time permitted.

Because they were designed to interconnect headquarters, early (first-generation) trunk networks followed the chain of command, which required headquarters to act as both tactical bases and large communications nodes. This dual role caused significant conflict in siting and hampered the mobility of commanders, as any movement of the headquarters disrupted communications.

Second-generation networks alleviated the siting quandary of the headquarters by separating the communications and tactical responsibilities through the provision of a separate communications center that could be sited separately from the headquarters. However, the communications center had to shut down to move with the headquarters, which limited command and control in fast-moving tactical environments.

Third-generation networks provided a headquarters with two communications centers, which allowed the headquarters to step up. A step-up element could move to the new location and set up one of the communications centers. Once this headquarters element was established and had communications with the old location as well as all subordinate commanders, command could be transferred and the elements at the old headquarters location could close down and move to join the new location. A headquarters was also more robust through having two communications centers, which allowed it to have alternative routes for communications in and out.

The logical extension of these developments is the fourth-generation, meshed, area trunk network. Most modern trunk communications systems have been developed as fourth-generation networks, for example: United Kingdom's PTARMIGAN; United States' MSE; Germany's AUTOKO; The Netherlands' ZODIAC; Italy's MIDAS; and France and Belgium's RITA [3].

2.5.1 Fourth-Generation Trunk Networks

Within a fourth-generation meshed trunk communications system, there are three basic entities: *trunk nodes* to provide backbone switches (we use the term "switch" here in a generic sense as a label for a device that enables interconnection between users and user applications); *bearers* to interconnect nodes into a network; and *access nodes* that serve a community of users. As illustrated in Figure 2.3, the trunk nodes are deployed across the corps area, providing a meshed network within which users can maneuver. While the network is deployed with a logical grid, connectivity will have varying physical configurations depending on the location of nodes as dictated by the terrain and tactical situation.

2.5.2 Trunk Nodes

Typically there are approximately 40 trunk nodes in a corps network. Nodes are normally allocated on the basis of four per division and the rest are deployed as corps assets. In principle, one element of the user community should be able to connect to the communications system at any point on the battlefield and have seamless access to any other user similarly connected, without any knowledge of their location. Similarly, while the network of

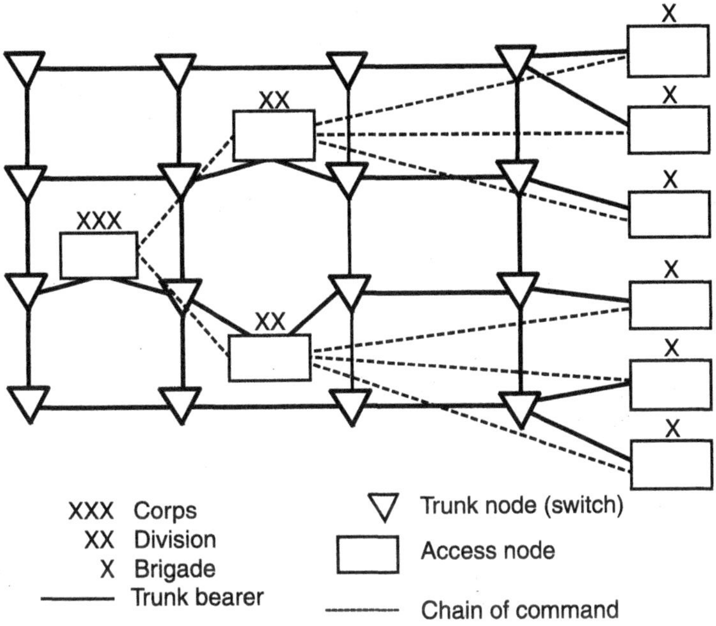

Figure 2.3 Architecture of tactical trunk communications systems.

trunk nodes may need to reconfigure to account for movements by user communities, any movement by the network should not have an impact on the service provided to user elements.

This separation of the user community and the switch network is often depicted as a "cloud" (see Figure 2.4) indicating that the topology and location of nodes within the network are largely irrelevant to the user community. User communities need only be able to access one of the trunk nodes in the network, through which communications can be achieved to any other group of users that is similarly connected.

As outlined earlier, in the trunk networks of most major armies, brigade headquarters is the lowest level of access node in the trunk network. The lowest extent of the trunk network is to battalion headquarters via mobile access. However, for most modern deployments using single-channel radio, the trunk network must extend to lower levels. While the trunk network supports communications between command posts from joint force level to brigade level, the major tactical need in most deployments exists below brigade—brigade headquarters, battalion headquarters, logistics installations, and headquarters of other major units in brigade areas of operations. However, there are a number of attendant problems with the provision of trunk facilities

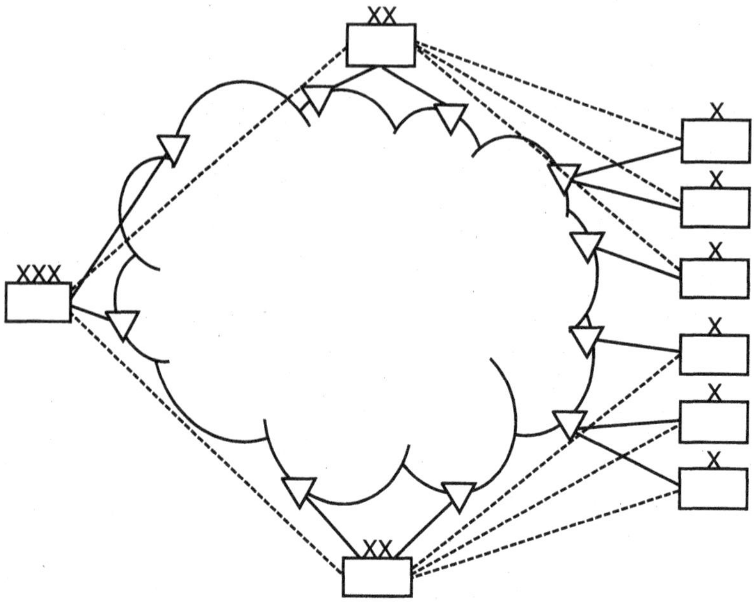

Figure 2.4 The fourth generation of trunk network represented as a "cloud" diagram.

to the small, mobile headquarters at battalion and below. While high capacity access is desirable at all levels, the mobility and tactical movement of low-level headquarters will tend to militate against the provision of large, vehicle-mounted equipment within the headquarters. At these levels, however, mobile access can be provided without restricting the tactical movement of commanders.

Fourth-generation networks provide great flexibility to a deployed force. Nodes can be deployed and redeployed to account for tactical movement by the force, and the network density can be adjusted according to force composition and distribution. Figure 2.5 illustrates the basic components of a trunk node.

The switch is the heart of the node providing a processor-controlled digital switch, which until recently has generally been based on an automatic circuit switch that has an embedded packet switch. However, most networks are investigating the incorporation of asynchronous transfer mode (ATM) switches or IP routers to accommodate traffic with varying quality of service requirements.

The node will also contain an operator interface to assist in engineering the switch's role in the network, manage the trunk encryption equipment, and allow some patching. This *node operation center* (NOC) can either be

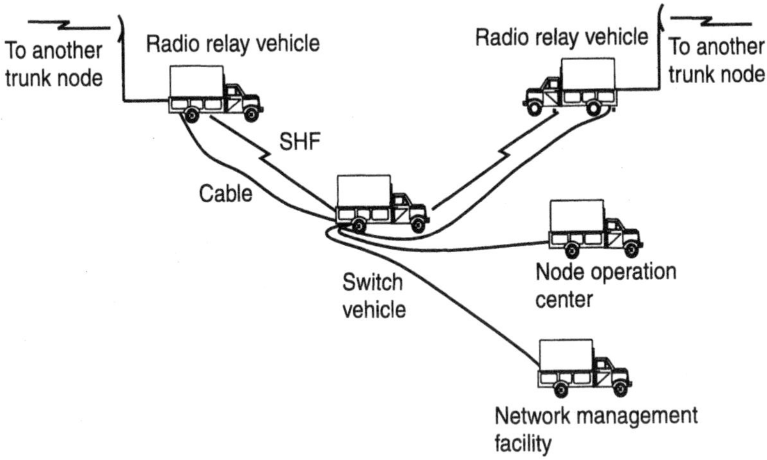

Figure 2.5 The basic components of a typical trunk node.

located in the switch vehicle or in a separate vehicle. Also collocated with the switch vehicle will be a *network management facility* (NMF), which performs link management of the radio relay links connected to the switch, as well as link engineering and frequency management.

Typically, a node will also include four or five radio relay detachments, each of which can terminate approximately three other radio relay detachments. The radio relay vehicles are often located on higher features surrounding the switch vehicle and are connected to the switch via cable or a "down-the-hill" SHF radio link.

2.5.3 Bearers

The interconnection of headquarters requires high-bandwidth digital communications that can be provided by any high-capacity bearers. HF communications are unsuitable for this requirement because of the low bandwidth and poor quality. Optical fibers are unsuited to mobile wide area networking and are only able to be used in short runs due to the time taken to lay and recover line. Satellite communications cannot be provided as part of the organic minimum-essential communications system because, in most armies, the satellite cannot be considered organic to the tactical formation. However, the use of satellite communications in the overlaid communications system is very useful to extend ranges of internodal links when required in dispersed operations.

Troposcatter communications do not have the setup and tear-down times required to support the deployment of trunk nodes. However, troposcatter communications can provide high bandwidths (up to 2 Mbps) over relatively long ranges compared to terrestrial systems (150–300 km for tactical systems). The utility of troposcatter is a useful adjunct to the organic assets of the tactical communications system. While troposcatter systems do not necessarily have to be provided on a scale to cover links between all trunk nodes, a small number of troposcatter terminals should be able to be provided within each tactical formation so that at least one troposcatter link can be established, as well as an additional terminal to anchor a link to a higher formation.

The most suitable bearer for the organic minimum-essential trunk communications is VHF/UHF terrestrial radio relay. High capacity links require high-gain (approximately 9 dB) antennas mounted on high (10m) masts. Even so, the range of these systems is limited to line-of-sight and needs to be extended for dispersed operations by the organic tactical airborne subsystem; organic troposcatter communications; and satellite communications provided by an overlaid communications system.

Although UHF and SHF frequencies provide sufficient bandwidth, they have the planning difficulty of requiring line-of-sight between antennas. Additionally, to maintain the required signal-to-noise ratios, particularly for data links, radio paths are limited to planning ranges on the order of 20 to 30 km. Often, therefore, two trunk nodes cannot be connected by one link, and *radio relay* is required through the insertion of an intermediate relay station in the link. This relay is sited so that there is a good line-of-sight path to both radio relay terminals, and the signals received from each terminal are automatically retransmitted to the other. Where long internodal links are required to support widely dispersed forces, there may be several relay stations.

As shown in Figure 2.6, where ranges are too long for the available terrestrial trunk assets, bearers may be replaced by satellite bearers or the range of terrestrial bearers may be extended by repeating through the tactical airborne subsystem.

2.5.4 Access Nodes

Subscribers gain access to trunk facilities through an access node, which is a processor-controlled switch capable of interconnecting all local subscribers as well as providing them with trunk access. In almost all networks, at least two levels of access node are provided: small access nodes for brigade

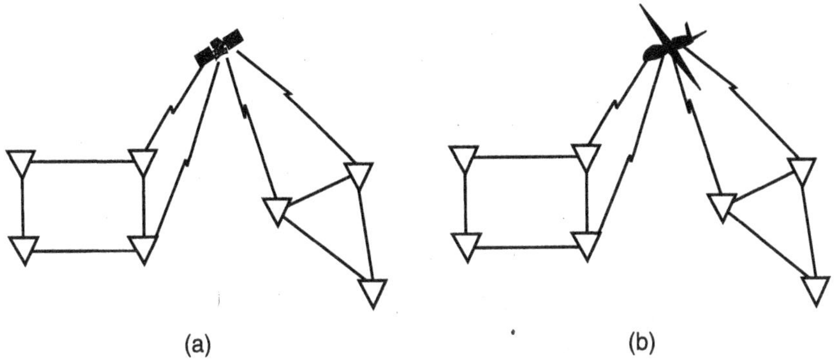

Figure 2.6 Range extension using (a) satellite internodal links and (b) the tactical airborne subsystem.

headquarters and large access nodes for divisional and higher level headquarters. Some networks provide access nodes to lower levels—at least to regimental or battalion command posts. In most networks, however, access to these levels is provided through the mobile access systems described next.

Access nodes are normally connected to the trunk network through radio relay bearers. However, in most networks, nodes can also be connected to overlaid systems such as commercial carriers and satellite or troposcatter bearers. Large access nodes generally have two connections to the network; small access nodes have one operational link and one engineered in standby in case of failure. Small access nodes normally provide for approximately 25 subscribers, while large access nodes connect about 150 subscribers. Figure 2.7 shows a generic configuration of an access node.

2.5.5 Mobile Trunk Access

Mobile remote access is required for trunk network subscribers. This functionality should include full duplex voice telephony as well as substantial data connectivity. In most major trunk networks this access is called *single channel radio access* (SCRA). It is normally provided by *radio access points* (RAP) that are connected to trunk nodes, as illustrated in Figure 2.8.

Mobile subscribers have secure duplex VHF radio access to the RAP and have voice, data, telegraph, and facsimile facilities similar to those available to static subscribers of an access network. The RAP can accept approximately 50 mobile subscribers affiliated within a 15-km radius. However, only approximately one quarter of these can make a simultaneous call. RAPs are deployed to provide overlapping coverage of the battlefield, in a manner similar to

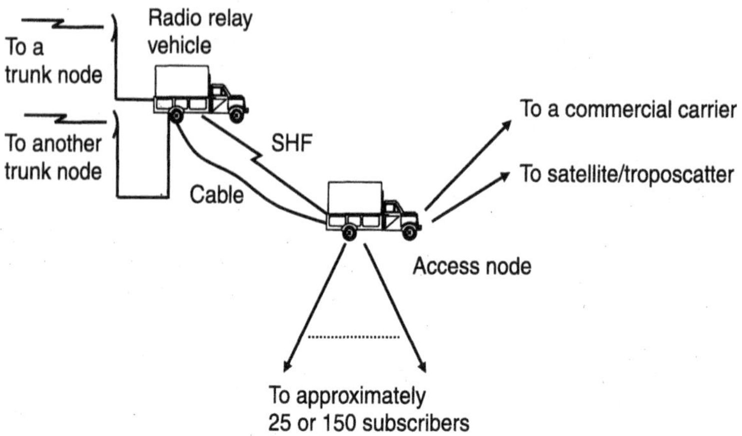

Figure 2.7 Generic layout of an access node.

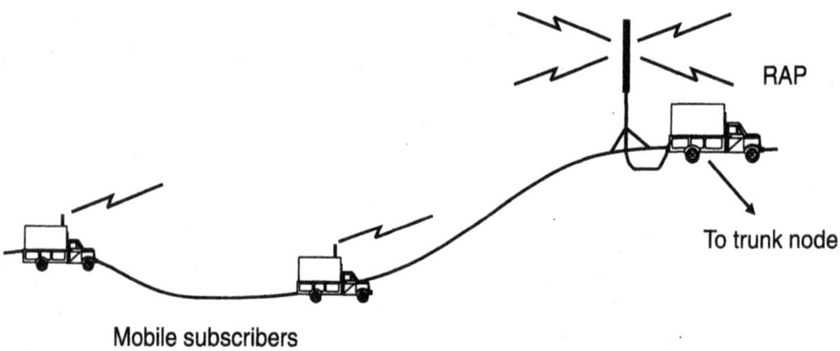

Figure 2.8 Configuration of the RAP for mobile subscribers.

that used to plan commercial cellular telephone networks. Variable power settings allow the RAP to reduce its RF signature.

A mobile subscriber is given a unique identification number during a process called affiliation, through which the subscriber is recognized and the subscriber's identity is validated. The network normally requires deliberate action on behalf of the user to affiliate. Some networks provide automatic handover of subscribers between RAPs; others require the subscriber to reaffiliate manually to the new RAP. Automatic power control is essential at the terminal to ensure similar power levels at the RAP from all subscribers.

2.5.6 Combat Net Radio Interface (CNRI)

In addition to the mobile trunk access provided to SCRA subscribers, most networks also provide an additional form of access to CNR users who are able to use their CNR to gain temporary (albeit limited) access to the trunk network. A CNRI vehicle provides semiautomatic access for several VHF and HF users. The users tune their CNR to the CNRI hailing frequency and arrange with the operator to have a call placed to a trunk subscriber. Calls made from the network are automatic and do not need operator assistance. As CNR nets are single-frequency and half-duplex, CNRI only provides rudimentary access to network facilities. In most current trunk networks, CNRI provides the only direct interface between the trunk and CNR subsystems. A much more sophisticated and ubiquitous interface is required if the CNR and trunk communications subsystems are to be seamlessly integrated.

2.5.7 Interfaces to the Tactical Trunk Subsystem

The current limited interface provided by CNRI is not sufficient. The tactical trunk subsystem is required to interface to other components of the tactical communications system as well as to the strategic communications system and to overlaid communications systems. These generic interfaces are illustrated in Figure 2.9.

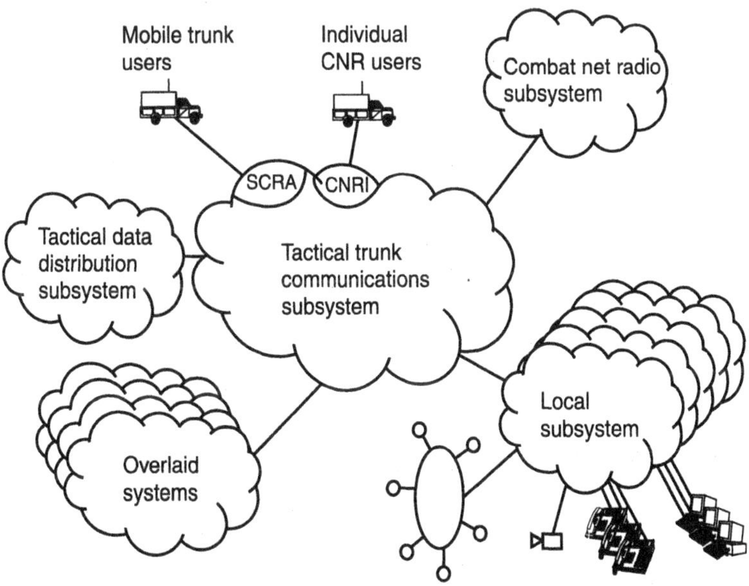

Figure 2.9 Tactical trunk communications subsystem interfaces.

2.6 CNR Subsystem

Military requirements for communications with fully mobile infrastructure have been met traditionally by CNR, which is the primary means of exercising command and control at brigade level and below. CNR combines the advantages of simplicity and flexibility with the ability to provide the all-informed communications that are essential for the close coordination of all-arms tactics in mobile operations.

The use of radio on the battlefield began in World War I as an alternative to the use of line as part of trunk communications. Examples included the connection of observation posts to artillery batteries, which avoided the laying of hundreds of miles of cable to support major offensives. Radio sets (and particularly their antenna systems) were initially too large to be of any great use to the infantry. However, as sets and antennas reduced in size, they began to be employed to form artillery-infantry nets and infantry-armor nets and became more useful in mobile operations.

By the end of World War II, CNR had become an important means of communications for infantry and other arms. Technical developments since that time have been evolutionary rather than revolutionary, and the tactical use of CNR has remained largely unchanged. The major difference is in the ability to pass data, although most CNRs are still analog radios and are not well placed to cope with the expansion in the volume of data expected in the next few years.

CNR is easily deployed and requires no infrastructure support. It is therefore likely to retain its place at unit level for the foreseeable future. At formation level, it continues to supplement the array of trunk communications systems available to the commander and his staff.

2.6.1 Key Architectural Issues for CNR

Some of the key architectural issues for CNR include the following.

Command and Control on the Move

User terminals and network infrastructure must be capable of operation while on the move without stopping. This requires that either there is no ground-based infrastructure or that this infrastructure is fully mobile. Command and control on the move makes the use of line unsuitable as a sole means of communications; it favors, however, the use of radio with omnidirectional antennas. Radios and terminals must also be small and robust with low-power requirements if they are to provide integral support to motorized and mechanized forces and especially light-scale forces.

Support for the Chain of Command

Support for the chain of command requires that communications between a commander and subordinates be attained with maximum efficiency. This is achieved in current systems by hierarchical nets that follow the chain of command. Support for the chain of command also requires that ground-based equipment used for communications within a unit or subunit is integral to that unit or subunit. The CNR subsystem should not constrain the locations of headquarters or other elements. This applies both while on the move and in static locations. A commander should also be able to alter command arrangements within a formation or unit without having to fundamentally restructure the CNR nets.

Voice and Data

Despite the need for voice remaining for the foreseeable future, the requirement to be able to send data will increase dramatically in the next few years. Both voice and data communications are required, therefore, although not necessarily at the same time or using the same equipment. Ideally, voice and data communications should be seamlessly integrated, using a single piece of equipment. A major disadvantage of separate equipment for voice and data may be additional weight.

Multiple Access

The spectrum available for military use is not likely to expand, while the number of systems that make use of the electromagnetic spectrum increases constantly. Sharing of the electromagnetic spectrum between users is required. Possible multiple access techniques [4] include frequency division multiple access (FDMA), synchronous time-division multiple access (TDMA), carrier-sense multiple access (CSMA), and code-division multiple access (CDMA)—see Table 2.1. Multiple access may be provided using a combination of these basic techniques. In current systems, multiple access is achieved by grouping stations into nets. Each net operates in a single-frequency, half-duplex mode, with different nets being assigned different frequencies. For conventional CNR, each net operates on a single frequency, with FDMA being used to share the electromagnetic spectrum between nets. Within a net, a form of CSMA is used to share the channel capacity between stations on the net. For voice networks, CSMA takes the form of voice procedure, RATEL.

Multicast Capability

Traditional CNR provides an all-informed voice capability that is ideally suited to the coordination of all-arms tactics. This requirement for all-

Table 2.1
Multiple Access Techniques

Technique	Description
FDMA	In FDMA, a portion of the electromagnetic spectrum is allocated to each transmitter,, which can transmit in its allocated channel all the time. Allocation of frequencies can be fixed or on demand. The advantages of FDMA are that no central control station is required unless capacity is demand-allocated, close to 100% of the available channel capacity can be used, there is no loss of data due to one station overtransmitting another, and there is no delay introduced by the channel. The disadvantages of FDMA are that the allocation of capacity between transmitters is relatively inflexible and it is relatively difficult for one station to receive data from more than one transmitter.
TDMA	In synchronous TDMA, a fixed-length, periodic time slot is allocated to each transmitter. Timing synchronization requires either a central control station or regular transmissions from all stations. Each transmitter has available the whole channel capacity during its time slot, and must remain silent at all other times. Allocation of time slots can be by fixed assignment or on demand. The advantages of synchronous TDMA are that it is relatively easy for one station to monitor transmissions in all time slots, it is possible to use close to 100% of the available channel capacity if there is either a single transmitter or single receiver, there is no loss of data due to one station overtransmitting another, and there is a fixed upper bound on delay. The disadvantages are that timing synchronization between stations is required, guard intervals reduce channel capacity when multiple transmitters and multiple receivers are used, and allocation of capacity between transmitters is relatively inflexible.

informed voice capability is not likely to decrease. With the introduction of a data capability into the CNR subsystem, a corresponding multicast capability for data is also required.

Seamless Connectivity

Seamless connectivity should be provided both within the CNR subsystem and between this system and trunk subsystem. This may require the rebroadcast of data within the CNR subsystem or carriage by the trunk subsystem of some data whose source and destination both lie in the CNR subsystem. In practice, seamless connectivity is a higher priority for data than for voice

Table 2.1 *(continued)*
Multiple Access Techniques

Technique	Description
CSMA	CSMA techniques are a form of asynchronous TDMA in which there are no fixed time slots. A station wishing to transmit checks first that no other station is currently transmitting. If the channel is free, the station transmits; if not, it waits a random period of time and tries again. If both stations inadvertently transmit simultaneously, both recognize the collision, cease transmission, and wait a random period of time before trying again to access the channel. The advantages of CSMA are that it is relatively easy for one station to monitor all transmissions on the channel, no central control station is required, and the allocation of channel capacity is very flexible. For the transmission of data, the disadvantages of CSMA are that the best throughput that can be achieved is approximately 50% of the available channel capacity, data is lost due to one station overtransmitting another, and there is no fixed upper bound on delay. In VHF radios, such as the SINCGARS ASIP [5], where the total available data rate is approximately 16 Kbps, the throughput has been found to lie between 1 and 3 Kbps.
CDMA	CDMA allows a wideband channel to be shared by a number of narrowband sources by spreading their transmissions over the whole band. By using a different spreading sequence for each transmitter, multiple access is achieved. The advantages of CDMA are that no central control station is required unless capacity is demand-allocated, close to 100% of the available channel capacity can be used, there is no loss of data due to one station overtransmitting another, and there is no delay introduced by the channel. The disadvantages of CDMA are that the near-far effect makes it infeasible to have more than one transmitter and more than one receiver operating simultaneously, the allocation of capacity between transmitters is relatively inflexible, and it is relatively difficult for one station to receive data from more than one transmitter.

and, in fact, may be feasible only for data. Typically, the interface to the trunk subsystem would be part of that subsystem.

Security

Secure communications should be provided at all levels. Secure operation for voice and data can be provided by external cryptographic equipment; more recent designs have an integral cryptographic capability.

Electronic Protection

EP is required to provide LPI and resistance to jamming. This might include passive EP, such as terrain shielding or the use of directional antennas, and active EP, including direct-sequence spread spectrum and frequency hopping.

Capacity

In current systems, separate capacity is allocated for each CNR net. The minimum capacity is one voice channel per net, which may be time-division multiplexed, for example, using CSMA between a number of terminals. Future requirements will not reduce the need for voice capacity for command and control, but may impose an additional requirement for data, some of which will replace certain voice traffic.

2.6.2 Range of Communications

Within the constraints of terrain limitations, the frequency and power of transmission are key factors that control the range over which communications is possible. The main frequency bands used for CNR are HF and VHF. UHF is used in some applications where short-range, line-of-sight communications are possible.

In the HF band, military communications occur in the frequency range from 2 to 30 MHz. Within this band the range of communications is frequency-dependent and two modes of propagation are possible:

- *Sky wave.* Sky-wave communications rely on transmitted energy being refracted by the ionosphere. Communications over very long ranges are possible. For given transmit and receive locations at a given time, sky-wave communications will only be possible within a limited part of the HF band. HF sky-wave antennas tend to be large and not suited to communications on the move.
- *Surface wave.* Surface-wave propagation uses the Earth's surface as the medium of propagation. Effective surface-wave propagation requires a conductive surface, such as salt water or wet, fertile ground. In optimal conditions, the maximum range may exceed 100 km. HF surface-wave propagation is always vertically polarized, allowing communications on the move with whip antennas.

CNR in the VHF band usually operates between 30 and 88 MHz. Operation at higher frequencies is usually prevented by commercial use of these bands. VHF propagation in this band is terrain limited to radio line-of-sight, which exceeds optical line-of-sight by approximately 15%. Increasing transmission power will not significantly increase the range. Range extension is better provided by using an elevated antenna or by retransmitting between nets.

2.6.3 Modern CNR Solutions

In summary, traditional CNR provides a single voice channel. Multiple access is provided within nets by CSMA, and between nets by FDMA. CNR is used in the HF (2–30 MHz), VHF (30–88 MHz), and UHF bands. The modulation scheme for HF CNR is usually single sideband (SSB), with a channel bandwidth of 3 kHz; for VHF, frequency modulation (FM) is commonly used with a channel bandwidth of 25 kHz or 50 kHz. However, this configuration is not sufficient as a basis for the development of a CNR subsystem required to support future land warfare.

This section describes the two candidate solutions for the provision of a fully mobile tactical communications system: CNR with a capability to pass digital data (data-capable CNR), and packet radio.

2.6.3.1 Data-Capable CNR

Data modems have been available for CNR for more than 20 years. A single HF CNR channel can carry up to approximately 2,400 bps, while a VHF or UHF CNR channel can carry 16 Kbps with a channel bandwidth of 25 kHz. CSMA is used to control multiple access. Where automatic control is provided, an operator can prepare a message and have this message transmitted asynchronously.

Voice and data can be supported with a single equipment. This is common for VHF CNR, although a separate modem is often required for carrying data on HF CNR. A net can operate in either a data mode or voice mode. A data net may drop back to voice operation when an operator presses the pressel switch. Switching or routing of data requires external equipment in addition to the data-capable CNR.

2.6.3.2 Packet Radio CNR

The major disadvantage of data transfer over CNR is that there is no support for automatic rebroadcast of data in fragmented nets. This can be overcome, at the expense of added complexity, by the use of packet radio. Both voice and data communications can be supported by means of a digital internetted radio. Examples of military use of this technology include the BOWMAN [6] HF and VHF radios currently being procured for U.K. land forces and the U.S. SINCGARS ASIP.

The data mode is the default mode of operation. In this mode, internet bridges allow automatic delivery of messages addressed outside a net. These bridges may be located at any point, but could be expected to be located in headquarters where stations currently exist on two adjacent nets. This is illustrated in Figure 2.10.

In the data mode, messages are automatically relayed if a net becomes fragmented. This process is known as *intranet rebroadcast.* In reality, internet rebroadcast may be limited to two or three hops. The routing/multiple access problems of such a fragmented net have yet to be addressed in detail in new systems such as the British BOWMAN network.

The secure-voice mode has priority over the data mode, and is entered immediately when an operator depresses the pressel switch. In the voice mode, operation of nets is identical to that of conventional secure-voice radio nets. The data mode is resumed immediately when transmission of voice ceases. Voice operations in fragmented nets are similar to that of conventional CNR nets.

Advantages of the tactical *inter*net are that data can be passed automatically between any two locations on the battlefield, without manual retransmission at net boundaries; the support of external network devices is not required; and messages can be routed around failed parts of the network, because the transmission path is not fixed. Disadvantages of the tactical internet are that, because of the low data rates available (1–3 Kbps), the network can easily become congested; and it is probably necessary to limit the number of

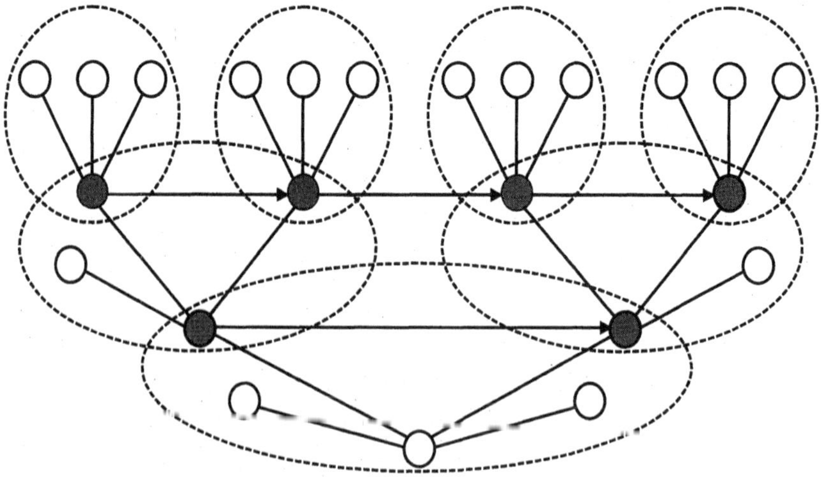

Figure 2.10 CNR tactical internet.

intranets across which a message is passed to prevent the whole network from becoming congested in the event of failure of one part of it.

Advantages of the tactical *intra*net are that the efficient operation of a fragmented net is possible; rebroadcast is provided with a single radio at each site; enemy intercept of transmissions is hindered by use of lower power levels than would be required for direct communication between all stations on a net; and traffic analysis of a net may be made more difficult, with many transmissions being rebroadcasts rather than new messages. Disadvantages of the tactical intranet are that an operator has less control over the transmission from his or her radio than in traditional hierarchical nets; because of the low data rates available and the inefficiencies of multiple-access protocols, the network can easily become congested; and operation with highly-fragmented nets does not appear to be feasible, with intranet rebroadcast limited in practice to one or two hops.

2.7 Tactical Data Distribution Subsystem

The requirement for providing real-time situational awareness creates two key difficulties for the CNR subsystem. The first of these difficulties is the capacity available in the CNR subsystem, which is unlikely to exceed 1 to 3 Kbps on any one net. The capacity required for real-time situational awareness in a mechanized brigade may, however, exceed 500 Kbps. A foot-mounted infantry brigade will have a much lower data requirement due to the slower pace at which it can move. The second difficulty is that aggregation of data at each level of the chain of command tends to cause an increasing data requirement at the next higher level.

The tactical data distribution subsystem addresses these two difficulties, providing a high-capacity, homogeneous, wireless, data network across the brigade area of operations. The tactical data distribution subsystem does not need to support voice, which is already well supported at high levels by the tactical trunk subsystem and at low levels by the CNR subsystem. Indeed, support for voice at low levels would require very high data rates in the tactical data distribution subsystem, possibly in excess of 1 Mbps for a brigade-sized force. These rates are very difficult to achieve.

While its main purpose is the efficient carriage of data for real-time situational awareness, the tactical data distribution subsystem can also be used to carry other types of data, such as reports and returns.

A repeated TDMA network overcomes the inefficiency of packet radio by requiring all stations to transmit in allocated time slots. All network timing and operation are controlled by a central station. A number of repeated TDMA radios are currently in service in the United States, including the Enhanced Position Locating and Reporting System (EPLRS) radio and TADIL-J (Link 16) [7]. The suitability of repeated TDMA networks is described here with reference to the properties of the EPLRS radio. The additional noncommunications services provided by the EPLRS system, such as position locating and reporting, will not be discussed.

The EPLRS radio supports a variety of data communications services, providing both point-to-point links and an extensive multicast capability, including all-informed nets. Data rates up to 57,600 bps per connection, known as a *needline,* are possible. Common use sees one EPLRS user community per brigade, with a maximum area of approximately 47 × 47 km. Each EPLRS user community has a maximum practical data capacity of between 300 to 450 Kbps, depending on configuration. This capacity is reduced when retransmission is required because multiple user communities are operating in the same area. Capacity is also reduced when the area over which the user community operates increases, because of the requirement for larger guard intervals between TDMA slots.

Multiple access is achieved using a combination of FDMA, TDMA, and CDMA. The EPLRS net control station (NCS), which is vehicle-mounted, controls the net, and provides timing for synchronization. All other EPLRS stations can be either vehicle-mounted or man-packed. Retransmission is used to extend the area of the EPLRS net beyond line-of sight. Frequencies in the range from 420 to 450 MHz are used, with the band being segmented into eight channels.

Resistance to jamming is provided through direct-sequence spread spectrum transmission; frequency hopping operation; error detection and correction; and network management that facilitates the automatic routing and rerouting of messages in the EPLRS network, using any user unit as a relay of opportunity. The combination of direct-sequence spread spectrum and frequency hopping also provides LPI. Security is provided by an embedded cryptographic system.

The network management capability also enables EPLRS to automatically build the network from scratch with no prior connectivity information and to automatically adapt to the changing battlefield conditions of terrain masking, user motion, jammer dynamics, and varying subscriber data communication requirements. In addition, the net management design accommodates the assignment and deletion of military users to the network.

Continuity of operations in EPLRS is maintained by software that permits data communications to continue along established paths if an NCS is lost. If the loss of a station occurs, the division's NCS or an adjacent brigade's NCS automatically assumes net control of the affected user community. Additional continuity of operations can be gained by the placement of an additional NCS in division rear to assume net control either in planned displacement or during unplanned sudden loss of any NCS. In dispersed operations a brigade may need an internal redundant NCS.

2.8 Tactical Airborne Subsystem

An airborne communications platform is required to provide additional capacity and to extend ranges of the CNR and tactical trunk subsystems when dictated by the operational scenario. Traditionally, the solutions to the provision of high-capacity, long-range communications have relied on the use of satellite-based services. However, subspace platforms (airborne, rather than spaceborne) offer a viable alternative, with the potential to deliver a broader range of services more cost-effectively.

The tactical airborne subsystem provides an airborne platform that carries robust communications packages to support command and control across wide areas. High-gain antennas coupled with the ability to loiter at high altitudes for extended periods will enable tactical users equipped with lightweight omnidirectional antennas and low-power radios to establish long-range communications from mobile platforms. This capability will provide a significant improvement in communications ranges and will enhance the ability for commanders to command and control on the move. The tactical airborne subsystem is required to provide range extension of CNR, tactical trunk, and tactical data distribution subsystems; additional communications services, including surrogate satellite communications; and coverage extension of overlaid communications systems, such as personal communications systems and the theater broadcast system.

As illustrated in Figure 2.11, the range extension offered by an airborne communications platform is very significant. In their normal terrestrial modes, CNR and radio relay are generally terrain-limited, not power-limited. For example, ground ranges of CNR are limited to 5 to 15 km depending on how high the operator can elevate the antenna. An airborne communications platform extends those ranges to up to 500 km, which is a dramatic improvement. For example, had an airborne communications platform been available during the Gulf War, true communications-on-the-move could have been provided to support a fast-moving, wide-ranging envelopment, at a time when terrestrial networks were stretched to the breaking point [8]. An

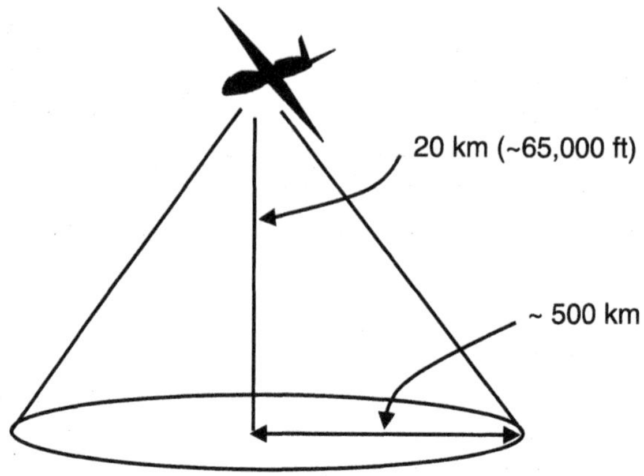

Figure 2.11 Footprint of airborne communications platform.

airborne communications platform would provide commanders with command and control on the move over a large operational area.

Further extension in coverage can be provided by cross-linking between platforms as illustrated in Figure 2.12. However, it should be noted that cross-linking requires high-gain antennas that have to be accurately pointed to another airborne platform. This will represent some engineering challenges for most platforms except those that can be maintained in a geostationary position, although modern phased-array antennas go some considerable way to solving this problem. The inclusion of a cross-linking capability will reduce the communications payloads on board the platform.

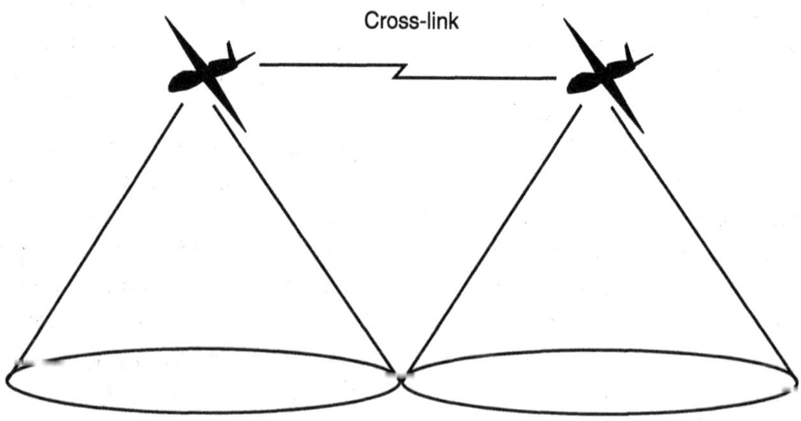

Figure 2.12 Increase in coverage by cross-linking between airborne platforms.

2.8.1 Additional Communications Services

In addition to range extension, a number of additional communications services can be provided from the airborne platform. An airborne platform presents the opportunity to provide true joint and combined communications in a simple manner. Figure 2.13 illustrates how the airborne platform can relay communications and establish a net between joint assets as well as provide reach-back communications by LEO or GEO satellite.

It may also be possible to mount a surrogate satellite transponder on board the airborne platform. In-service satellite ground terminals could be used to communicate to the surrogate transponder rather than to a satellite. Since the airborne platform is much closer, lower powers (higher data rates) are possible throughout the AO, without having to be within the satellite footprint.

It may be possible to mount a cellular PCS base station on the airborne platform and provide digital mobile telephony coverage within the footprint. However, commercial PCS base stations tend to be large and heavy, although there are some moves within the commercial industry to reduce the size and weight, since there a number of programs in the United States and Europe that are examining the possibility of airborne PCS base stations. The large power requirements of a base station provide an additional problem. How-

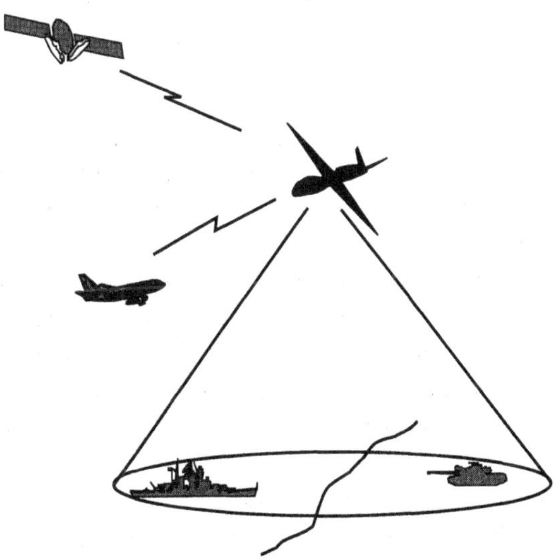

Figure 2.13 Improved battle space communications using the airborne communications platform.

ever, even if a base station is not included in the platform payload, the airborne platform has the potential to support the connection back to a base station on the ground.

The airborne platform also offers two opportunities with regard to theater broadcast:

- *Broadcast relay.* One of the difficulties in getting satellite theater broadcast to tactical units is the difficulty that combat troops have in deploying reasonably sized satellite antennas. Broadcast relay would take the satellite theater broadcast on the airborne platform and rebroadcast it into the footprint in the VHF/UHF band. Broadcast communications can then be provided to tactical users who can receive broadcasts with small omnidirectional antennas and less expensive receivers.
- *In-theater broadcast.* It is doubtful whether the brigade commander would allow theater broadcast into the brigade area without modification by the brigade staff; the most useful product for brigade units is a broadcast of the brigade/divisional view, rather than a strategic one. In that case, the airborne platform allows the brigade commander to take the satellite broadcast at brigade headquarters, modify and add information, and then provide an in-theater VHF/UHF broadcast through the airborne platform. Again, any mobile units in the footprint can receive high-capacity broadcast with small, omnidirectional (whip) antennas.

The deployment of an airborne communications platform opens up many other communications uses. For example, should commanders decide that videoconferencing is required, terrestrial solutions are difficult to arrange. However, an airborne platform provides an ideal vehicle for the delivery of videoconferencing facilities into the footprint. Also, high-capacity, long-range communications are normally difficult for special forces, who are forced to operate using HF. An airborne communications platform can extend high-capacity communications to special-forces patrols, which can be equipped with small, low-powered receivers with small, omnidirectional antennas.

In addition to communications uses, the tactical airborne subsystem may also potentially be used for noncommunications purposes. Just as the airborne platform provides an ideal communications base, it also provides an excellent platform for EW. Ground-based intercept is invariably terrain-limited, and many more ground assets would be required to have the same coverage as an airborne EW platform. The airborne platform could also provide broad area surveillance of the AO in a range of frequencies, through its ability to carry a range of optical, infrared, multispectral, and synthetic aperture radar (SAR) sensors. However, it is not likely that such additional uses could be incorporated on a communications platform without suffering some loss in communication ability due to the incorporation of extra equipment, antennas, and so on. Sensors also come with considerable additional space and weight requirements, as stabilization systems are often several times heavier than the sensors themselves. Additionally, the operation of a multirole platform may be difficult to coordinate if the other tasks demanded an operational profile at odds with its communications tasks. For those reasons, it is most likely that additional tasks such as EW and remote sensing would be conducted from separate dedicated platforms, although the same type of platform could be utilized in each role.

2.9 Potential Targets for Tactical Communications EW

In this chapter we have described the adversary communications systems that can be targeted by tactical communications EW. More explicitly, in subsequent chapters we discuss the following types of targets:

- CNR, VHF, and HF radio nets;
- Terrestrial trunk radio (HF, VHF, and UHF);
- SCRA;
- Radio links to airborne communications subsystems;
- Radio links for tactical data distribution subsystems;
- Interfaces between subsystems;
- Communications equipment including radio and line.

Overlaid systems such as satellite and PCS are also vulnerable to EW but would not normally be attacked by tactical assets due to the level of control required to ensure the coordination of monitoring and jamming missions on such systems. In general, tactical EW will only be employed to attack tactical communications systems. Where systems such as satellite and PCS systems have a broader impact and range than the tactical environment, EW is normally coordinated and conducted by operational or, more likely, strategic assets. However, as overlaid systems are deployed more commonly at the tactical level, it is possible that tactical assets will be targeted against such systems. The following chapters therefore also consider the issues associated with the attack of overlaid systems.

2.10 Summary

This chapter has described the tactical communications systems that are the target of tactical EW systems. The focus of the chapter has been on understanding the current structure of tactical communications systems as the basis for our discussion in subsequent chapters of methods of attack by tactical communications EW systems. The architecture developed illustrates the interrelationship of systems required to provide the battlefield network that supports operational concepts such as network-centric warfare. Current tactical trunk communications, combat net radio, tactical data distribution, and tactical airborne subsystems have been described in sufficient detail to allow an analysis of their strengths and weaknesses from an EW point of view. Chapter 8 discusses the future directions in which tactical communications systems are being developed.

Chapters 3, 4, and 5 now consider the EW functional areas of EP, ES, and EA as they apply to the tactical communications systems described here.

Endnotes

[1] Ryan, M., *Battlefield Command Systems*, London, U.K.: Brassey's, 2000.

[2] Ryan, M., and M. Frater, "An Architectural Framework for Modern Tactical Communications Systems," *IEEE Military Communications Conference (MILCOM 2000)*, Los Angeles, CA, October 23–25, 2000.

[3] Further information about trunk networks is contained in:
Hewish, M., "Tactical Area Communications Part 1: European Systems," *International Defense Review*, May 1990, pp. 523–526.
Hewish, M., "Tactical Area Communications Part 2: Non-European Systems," *International Defense Review*, June 1990, pp. 675–678.
Ryan, M., *Battlefield Command Systems*, London, U.K.: Brassey's, 2000.

[4] Sklar, B., *Digital Communications: Fundamentals and Applications*, Upper Saddle River, NJ: Prentice Hall, 2001.

[5] Kagan, M., "Redesigned Communication Equipment Strengthens First-to-Fight Operations," *Signal*, March 1999, pp. 33–36.

[6] Witt, M., "Britain's BOWMAN: Back to the Beginning," Military Technology, December 2000, pp. 64–68.

[7] Further information about JTIDS can be found in:
Toone, J., and S. Titmas, "Introduction to JTIDS," *Signal*, August 1987, pp. 55–59.
Stiglitz, M., "The Joint Tactical Information Distribution System," *Microwave Journal*, October 1987.
"JTIDS/TIES Consolidate Tactical Communications," *EW*, September/October 1977.
MIL-STD-6016, "DoD Interface Standard Tactical Digital Interface Link (TADIL) J Message Standard," February 1997.

[8] McAllister, M., and S. Zabradac, "High-Altitude-Endurance Unmanned Aerial Vehicles Pick Up Communications Node," *Army Communicator*, Spring 1996.

3

Electronic Protection

3.1 Introduction

Electronic protection (EP) comprises those actions taken to protect personnel, facilities, and equipment from any effects of friendly or adversary employment of EW that degrade, neutralize, or destroy friendly combat capability. In other words, EP is concerned with minimizing the effect of both friendly EA and an adversary's EA and ES. While EP is traditionally most concerned with protecting communications equipment, it is applicable to the protection of all systems [1].

EP is usually divided into *passive EP* and *active EP,* as shown in Figure 3.1. Passive EP comprises measures that are not detectable by an adversary, and is concerned with tactics and procedures for providing electronic protection, including terrain shielding. Active EP, whose measures are detectable by an adversary, is concerned with providing protection by the use of special equipment, or special operating modes of equipment.

One important way in which EP differs from the other EW subdivisions is that it should be practiced by all tactical units, not just by specialist EW units. Unlike other aspects of EW, EP is directly associated with the tactical communications system. The techniques discussed in this chapter relate to the employment of the tactical communications system or to specific features of the equipment that makes up the tactical communications system.

3.2 Passive Electronic Protection

Passive EP makes use of tactics and procedures to reduce the exposure of friendly use of the electromagnetic spectrum to both friendly and adversary

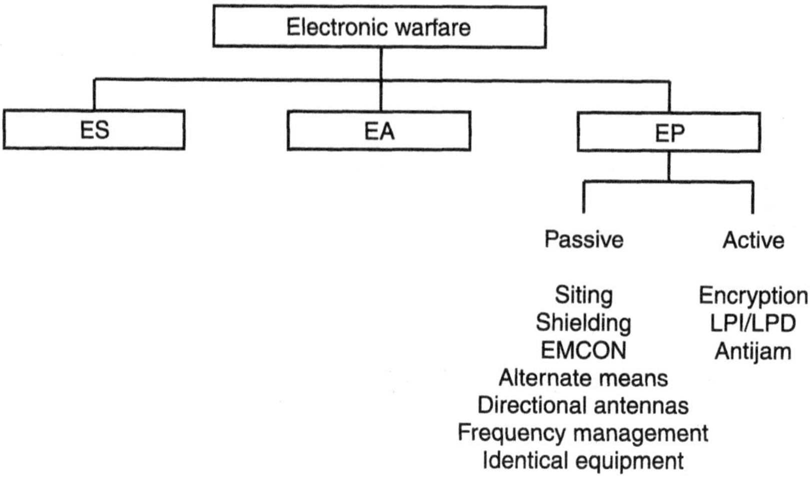

Figure 3.1 EW architecture.

EA. These tactics and procedures include the use whenever possible of identical equipment, shielding, emission control (EMCON), the use of directional antennas, frequency management, the provision and use of alternate means, and siting of communications.

3.2.1 Identical Equipment

Variations in the characteristics of transmissions from different types of equipment can be used to infer a variety of information about a force. In many armies, for example, older CNR equipment used by reserve forces uses a channel bandwidth of 50 kHz; newer CNR, used by higher-readiness troops, has a channel bandwidth of 25 kHz. The type of unit can therefore be inferred from the channel bandwidth being used by integral radio systems. Similarly, coalition forces can often be separated by their equipment. The use of identical equipment removes this source of information.

3.2.2 Shielding

All electronic equipment radiates electromagnetic energy and is potentially vulnerable to the effects of electromagnetic energy radiated by other electronic equipment. Shielding is a means of reducing both the amount of energy radiated and the vulnerability to received radiation, and offers protection against both adversary ES and EA.

The radiation generated by an electronic system is a potential target for adversary ES. While the potential range over which this ES is likely to be effective is small (perhaps hundreds of meters at most), ensuring that adversary ES is excluded from such an area may be difficult, especially in strategic systems and logistics installations. The use of shielding to reduce vulnerability to adversary ES is known as TEMPEST [2].

The use of shielding to protect systems from external radiation counters the potential use of neutralization as a form of EA. The major threat historically has been the EMP resulting from a detonation of a nuclear weapon. In the future, such high levels of radiation may be able to be generated by small, nonnuclear devices. This issue is explored further in Chapter 7.

3.2.3 Emission Control

An emission control plan may be used to reduce or alter the electronic signature of a force. This may be achieved by reducing the level of particular types of transmissions or the insertion of dummy traffic. Two commonly used modes of emission control are *radio silence,* in which all communications transmitters are deactivated, and *electronic silence,* in which all electronic emitters, including radars, are deactivated.

When planning emission control, it must be recognized that the imposition of radio and electronic silences may place operational constraints on the force, which may in turn increase rather than decrease its vulnerability.

3.2.4 Directional Antennas

Directional antennas are commonly used in tactical communications systems in which there are a single transmitter and a single receiver. An example of such a system is a radio relay link. The use of a directional antenna for transmission also maximizes the amount of power radiated toward the intended receiver and reduces the amount of power radiated in other directions. With careful siting, this can be used to minimize the power received at potential sites for an adversary's ES. The use of a directional antenna for transmission maximizes the received power from the intended transmitter, and reduces the amount of power received from other directions. This directionality can be used to reduce the effectiveness of an off-axis jammer.

For systems such as CNR, where there is more than one receiver, omnidirectional antennas are usually used. The use of directional antennas is therefore limited. However, some of the benefits of a directional antenna can be obtained by using an antenna with a steerable null that can be directed towards an adversary's ES or EA site. However, the practical use of null-steering antennas is restricted in highly mobile nets where the locations of friendly units change rapidly.

3.2.5 Frequency Management

Frequency management is required as part of communications planning to allocate the available capacity of the electromagnetic spectrum to users to avoid cochannel interference. This planning takes into account the relative locations of units, the likely range of communications (which is frequency-dependent), and interference that may be caused by harmonics and intermodulation products between systems located close together. Another important aspect of frequency management that can assist EP measures is the allocation of alternate frequencies that can be used in the event of interference or jamming.

3.2.6 Alternate Means

EP can be provided by the use of a variety of alternate communications means, which may be used either to overcome interfering EA or to reduce the susceptibility to ES. Some alternate means, such as line and messenger, do not involve radio. They are therefore particularly useful during periods of radio silence, but tend to limit mobility. Others may provide a different type of radio channel, such as using a trunk circuit rather than a CNR channel.

3.2.7 Siting

Planning for the siting of communications systems takes into account propagation of radio waves between chosen sites. This planning should also seek to minimize an adversary's potential use of EA and ES.

One means of doing this is to use *terrain shielding* (or *terrain screening*), as illustrated in Figures 3.2 and 3.3. In Figure 3.2, where transmitters are located on the tops of hills, reception is possible at a potential site for an adversary's ES or EA assets. However, by moving the transmitters down from the tops of the hills as illustrated in Figure 3.3, communication is maintained while denying coverage at the adversary's location.

Terrain shielding is most effective at VHF and higher frequencies where most tactical CNR radios operate. At these frequencies, propagation is essentially line-of-sight, and therefore tends to be terrain-limited rather than power-limited.

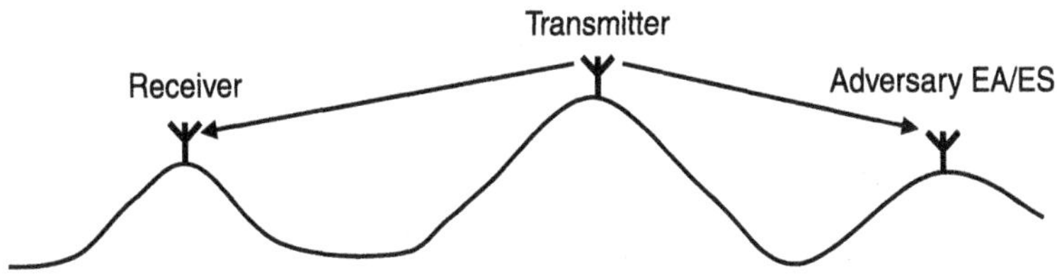

Figure 3.2 Ineffective terrain shielding.

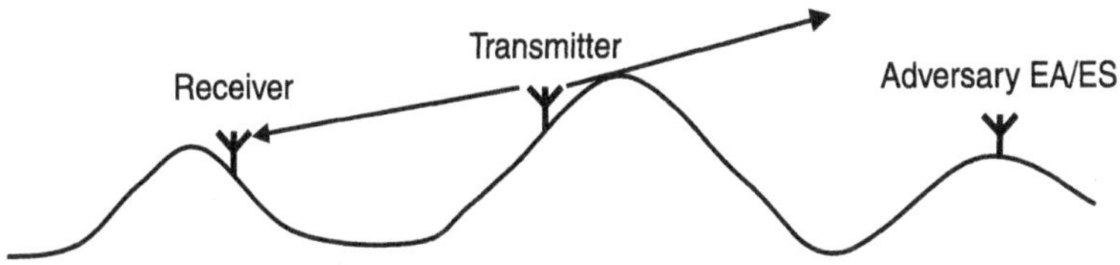

Figure 3.3 Effective terrain shielding.

Effective siting of radio equipment for EP requires a knowledge of the effective range of communications. This requires carrying out radio path planning not only for the communications network being planned, but also for potential sites for an adversary's ES and EA assets.

Some protection can also be gained from the careful use of shields in the area of the communications system, such as metal buildings, to provide screening. It is usually difficult, however, to predict the effectiveness of such procedures in advance. Although their effectiveness against an adversary's jamming may be measurable while the jammer is inactive, it is difficult to estimate their effectiveness against ES.

3.3 Active Electronic Protection

Active EP uses special equipment or operating modes of equipment to provide protection. The specific aims of protection provided by active EP are communications security, low probability of intercept (LPI), and resistance to jamming. Masking, which is the jamming of an adversary's ES receivers in such a way as to prevent their inhibiting friendly use of the electromagnetic spectrum, is sometimes also classified as EP. In this book, however, masking is covered under EA in Chapter 5.

A wide variety of techniques are used to provide active EP, including encryption, modulation, error-protection coding, burst transmissions, narrowband excision, diversity, free-channel search, and appropriate use of retransmission.

Encryption Encryption is used to modify transmitted data in such a way that it can only be decoded by the intended recipient. Encryption also protects against the insertion of dummy traffic by an adversary; some types of encryption also protect against imitative deception. Encryption is typically possible only in digital systems.

Modulation Type Some forms of modulation provide higher levels of protection than others. *Spread-spectrum communications* involves the spreading of a transmission across a wide band of frequencies. This can provide both LPI and a high-level of resistance to jamming. *Morse code* can often be used successfully on communications channels where voice and data communications are not possible. The primary disadvantages are the very low data rates obtained and the large cost of training operators and maintaining their skills.

Error-Protection Coding The addition of error protecting codes can increase the amount of interference that can be tolerated for a given quality of service.

Burst Transmissions An adversary's EA and ES can be hindered by reducing the length of transmissions. This may be achieved, for example, by using data rather than voice. If the length of a burst is sufficiently short, an adversary's intercept, DF, or responsive jammer may not be effective.

Narrowband Excision The impact of an interfering signal that has a much smaller bandwidth than a transmission can be minimized by excising the part of the spectrum containing the interfering signal before decoding. In digital communications systems, this technique is usually useful only when applied in conjunction with error-protection coding.

Diversity Protection against jamming and other forms of interference can be gained by transmitting data over more than one channel between transmitter and receiver. Diversity can be achieved in space, time, and frequency.

Free-Channel Search Conventional CNR uses a single channel for each net, with operators required to change frequency manually to overcome interference or jamming. Free-channel search allocates a group of frequencies to a group of nets. Radios on the net automatically identify a free channel when they wish to transmit, and then signal this choice of channel to other stations on the same net. Each transmission on the net may therefore use a different frequency. The potential drawback of such a system is that a jammer may be able to keep stations locked to a particular frequency by making it appear that this frequency is still in use by the net.

Appropriate Use of Retransmission Retransmission is often required for line-of-sight communications. A radio-relay network, for example, is built from point-to-point links between nodes. In CNR, retransmission is often required to overcome the effects of terrain. Care in siting, use of manual retransmission, and action on jamming are required to limit the vulnerability of retransmitted links to jamming. Rebroadcast stations are typically sited on the tops of terrain features, where they are most vulnerable to ES and EA. Siting in less exposed locations may reduce this vulnerability, while still permitting effective rebroadcast. An automatic rebroadcast will transmit not only voice and data, but also jamming signals. Manual control may be used so that jamming signals are not retransmitted. A rebroadcast station may continue to rebroadcast on a frequency even when other stations on the net have moved to an alternate frequency due to jamming. This increases an adversary's difficulty in evaluating the effectiveness of jamming.

3.3.1 Encryption

Encryption protects digital data by transforming the original data *(plaintext)* into a different form *(ciphertext)* that can be revealed without disclosing the original data [3]. The basic structure of a secure communications system employing encryption is shown in Figure 3.4. Encoding (known as encryption) and decoding (decryption) involve a security-related transformation. Secret information, which is distributed by a trusted third party, is usually used to modify this transformation so that the security of the system can be maintained even if the encoding and decoding algorithms are widely known. This secret information is often referred to as a *shared secret* or *key.*

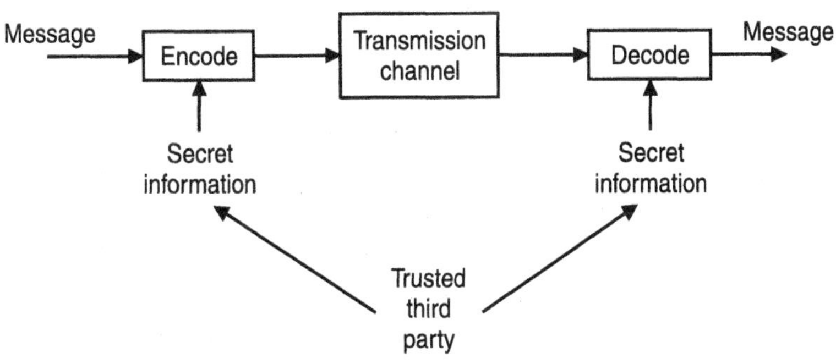

Figure 3.4 Generic structure of a secure communications system.

There are three main forms of encryption: bulk encryption, message encryption, and message-content encryption.

- *Bulk Encryption.* Here encryption of all data on a link on which transmission is continuous prevents both unauthorized reception and traffic analysis. This form of encryption is used by most trunk communications systems, and is sometimes referred to as *trunk encryption.* In addition to protecting against interception, this form of encryption protects against deception.
- *Message Encryption.* Individual messages on a link are encrypted, including both message header and contents. This form of encryption is used in most CNR systems. While the contents of the messages are protected from interception, message encryption does not prevent traffic analysis. If synchronization is achieved through a preamble, no protection is provided against deception by delayed replay of previous traffic.
- *Message-Content Encryption.* The bodies of messages are encrypted, leaving message headers in plain text. This form of encryption is often used in packet-switching systems. It has the advantage that intermediate switches and routers are not required to have attached cipher equipment. The disadvantage, however, is that detailed information that can be used in traffic analysis is provided in plain text. If synchronization is achieved through a preamble, this type of encryption does not protect against deception by delayed replay of previous traffic.

In general, the aim of an encryption system is to generate ciphertext that appears to be a random bit stream. In an ideal system, this randomness includes:

- *No structure observable in ciphertext.* Any structure of the plaintext, such as regular repeating patterns or different probabilities of occurrence for different symbols, should be removed in the encryption process to produce a ciphertext that has no regular structure.
- *Long key.* The maximum amount of work required for cryptanalysis is to decrypt the ciphertext using every possible key. The use of long keys maximizes the difficulty of this task. For a key that is n bits long, there are $2n$ possible keys.

- *Strong avalanche effect.* There should be large differences in ciphertext for similar keys with the same plaintext and similar plaintexts with the same key. Without the avalanche effect, similar plaintext can lead to similar ciphertext, greatly simplifying cryptanalysis.
- *Diffusion.* Every bit of plaintext should affect a large number of bits in the ciphertext.

In communications systems that introduce errors into transmitted bit streams, both the avalanche effect and diffusion can lead to a single bit error in the ciphertext that causes multiple bit errors in the deciphered plaintext. This phenomenon is known as *error extension.* For this reason, many practical encryption systems do not exhibit either a strong avalanche effect or diffusion.

In message encryption and message-content encryption, the encryption engine is restarted in a known state at regular intervals. In CNR, for example, where each transmission is independently encrypted, the decryption engine must know the initial state of the encryption engine at the start of each transmission. If this known state is always the same, the task of the cryptanalyst is greatly simplified. Some additional information that is known to both encryption and decryption systems is normally used to allow the encryption engine to bring each new message in a different state. Synchronization between transmitter and receiver can be achieved by transmitting a preamble containing the additional synchronization information at the beginning of each transmission, and possibly transmitting further synchronization information at regular intervals during the transmission; or deriving the additional synchronization information from an accurate time reference known to both transmitter and receiver.

3.3.1.1 Stream and Block Ciphers

A *stream cipher* operates on each bit of the transmitted message separately. This is usually implemented by generating a pseudorandom keystream. Each bit of the transmitted stream is the exclusive-or (XOR) of the corresponding bits of the keystream and the message, as shown in Figure 3.5.

A *block cipher* operates on blocks of the message. It requires buffering of a block of data, followed by processing and transmission of that block on the output channel.

Stream ciphers are easy to implement on synchronous communications channels because they produce one output bit for every input bit, produce only one bit-period of delay, and do not require additional buffering. Their disadvantage is that they cannot exhibit either diffusion or the avalanche effect with respect to the message.

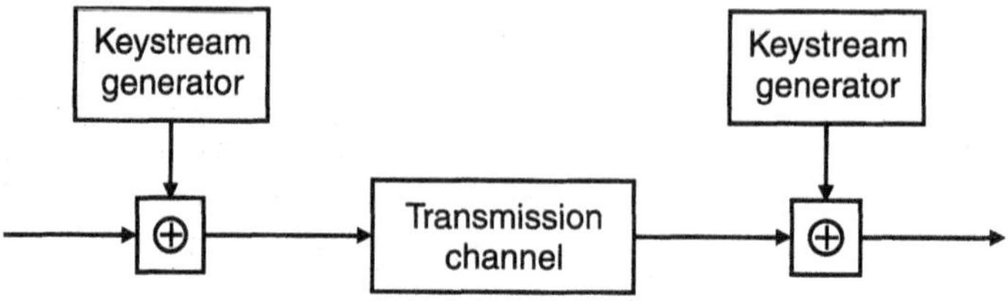

Figure 3.5 Structure of a stream-cipher system.

Many military encryption systems are used to encrypt data on synchronous communications channels. Such systems usually employ stream ciphers. For data transmission, this prevents the transmission of synchronization information, except as a preamble at the beginning of a transmission. Some voice systems steal bits during the transmission to transmit synchronization information, reducing their susceptibility to loss of synchronization when jammed. Off-line systems, such as one-time letter pads, are usually block ciphers.

3.3.1.2 Public-Key Versus Secret-Key Encryption

In traditional cryptosystems, the same key is used for encryption and decryption. This type of system is referred to as a *secret-key* encryption system, and is illustrated in Figure 3.6. The key that modifies the operation of the encryption and decryption engines is the shared secret.

Secret-key encryption can only provide limited security services. It can provide *confidentiality* (i.e., ensuring that messages are not disclosed to unauthorized parties). It is difficult, however, to guarantee the integrity of a message, because any party capable of decrypting a message is also capable of changing the message and reencrypting this corrupted version.

The solution to this problem is to separate the encryption and decryption keys, so that one can be made public without compromising the security of the other. This form of asymmetric encryption is known as public-

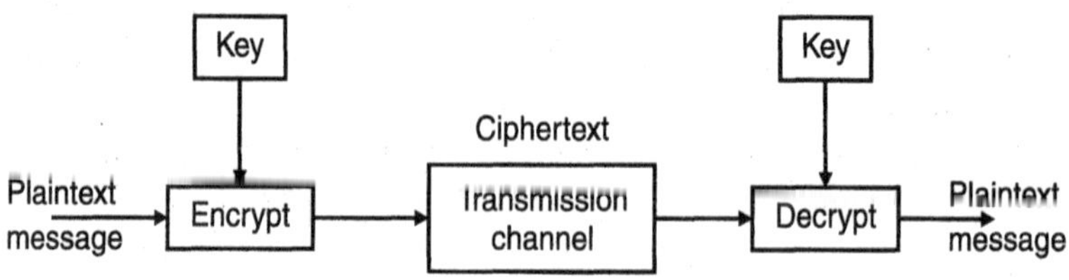

Figure 3.6 Structure of a secret-key secure communications system.

key encryption. Public-key encryption makes possible a range of additional security services, including integrity and authentication.

The technical basis of ***public-key encryption*** is in the differences in computational complexity for particular inverse mathematical operations. Multiplication, for example, is much less complex than factorization. Public-key encryption systems make use of the simpler operation to generate the keys. Any attack on the security of the system must be based on the more complex, inverse operation. It is therefore possible to create a system in which it is easy to generate a matched pair of encryption and decryption keys, but in which it also is very difficult to obtain one of these keys from the other.

The computational operations required for public-key encryption and decryption typically include multiplication and exponentiation. These operations tend to be best implemented using general-purpose computation engines, such as those found at the heart of a computer. Most secret-key systems utilize only basic logic operations (AND, OR, and XOR) and a small number of additions. These operations can be efficiently implemented in fast, special-purpose hardware. Because of this difference in the types of operations required, public-key encryption is much more expensive to implement, especially for high-speed communications.

Hybrid systems use secret-key encryption to transfer data, with the keys for the secret-key encryption being transferred using public-key encryption. Such systems have the advantage that they provide only small amounts of public-key ciphertext for cryptanalysis and favor a regular changing of the secret keys used to encrypt the data to be transferred.

Advantages of secret-key encryption over public-key encryption are simplicity of implementation and low computational complexity. The primary advantage of public-key encryption over secret-key encryption is the variety of security services that can be offered.

A number of secret-key encryption systems are in common use. The Data Encryption Standard (DES) has been established by the U.S. National Institute of Standards for the secure transmission of commercial data. It uses a 56-bit key to encrypt blocks of 64 bits of data. The short length of the key is a serious weakness, leading to the use of the so-called *triple-DES,* which provides a key length of 112 bits with backward compatibility with DES if the first and second 56-bit halves of the key are the same. The *International Data Encryption Algorithm* (IDEA) developed at the Swiss Federal Institute of Technology is also based on 64-bit blocks, but uses a 128-bit key. Skipjack [4] is a block cipher using a 64-bit block size with an 80-bit key, designed by the U.S. National Security Agency. This algorithm was initially classified, but has since been released publicly. The basic GSM A5 [5] is a stream encryption system that employs a 64-bit key. Because the keystream generator is restarted every 228 bits, however, it is effectively used as a block cipher.

The most commonly used public-key encryption systems are the Rivest-Shamir-Adleman (RSA) algorithm [6], and the Key Exchange Algorithm (KEA) [7].

3.3.1.3 Public-Key Authorities and Certificate Authorities

The establishment of secure communications requires a trusted intermediary to enable the secure distribution of public keys. Given the existence of this third party whose public key is already known to both parties—a *public-key authority* (PKA)—two parties A and B can exchange public keys and authenticate one another. For A to receive B's public key (Figure 3.7), A sends a plaintext message, which contains a time-stamp and a request for B's public key, to the PKA. The PKA sends A an encrypted message containing B's public key; A's original request, so that A can verify that it has not been altered during transmission from A to the PKA; and the original time stamp, so that A can verify that the message was generated in response to this particular request.

A uses the PKA's public key to decrypt this message. The placement of both A's original request and the original time stamp in the message returned by the PKA is required to prevent spoofing attacks on the key exchange.

The potential drawback of PKA is that it may become a bottleneck, especially if a large number of parties wish to establish secure communications. This problem is overcome by the use of a *certificate authority* (CA). The CA

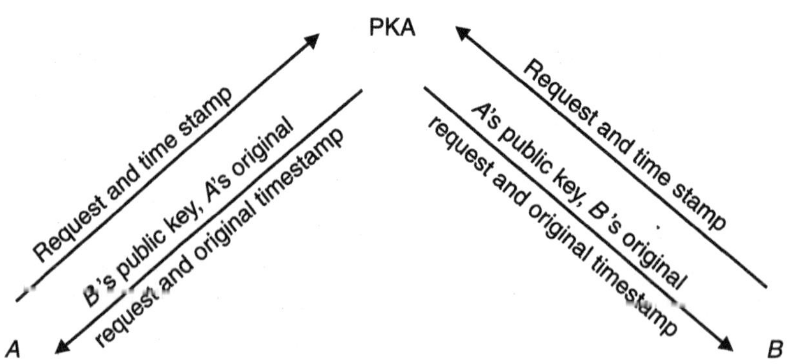

Figure 3.7 Use of the PKA.

creates certificates, which are encrypted messages containing the user's public key, the user's identification, and a time stamp.

The user then distributes this certificate with the time stamp. The authenticity of the certificate can be assured if the CA's public key is known. Furthermore, if the CA has suitable controls for the issuing of certificates, a certificate can authenticate the holder of the certificate.

3.3.1.4 Level of Security

The process of attempting to obtain plaintext without access to the shared secret is known as *cryptanalysis.* The computational resources required to attack a particular encryption algorithm define the difficulty of cryptanalysis and therefore the security of the algorithm.

An *unconditionally secure* encryption system is one for which it is impossible to obtain plaintext from ciphertext without possessing the shared secret. The only such system known is the *one-time pad,* which works by defining a sequence of randomly chosen transformations that are applied to each symbol of the plaintext to produce ciphertext. If a particular one-time pad is only used once, and the transformations are perfectly random, cryptanalysis is impossible.

However, most practical systems are not based on the one-time pad, and are therefore not unconditionally secure. While it is possible to apply cryptanalysis successfully to such systems, it may not be feasible. An encryption system may be *computationally secure* if either the cost of breaking the cipher exceeds the value of the encrypted information, or the time required to break the cipher exceeds the useful lifetime of the information.

Figure 3.8 shows how the time required for exhaustive cryptanalysis varies with the key length. The times shown take into account both processing time and Moore's law, which states that the computational power of computers increases by approximately a factor of 10 every five years. It can be seen that short keys provide only low levels of security. Longer keys, such as 128-bit and 256-bit keys, provide much higher levels of security, exceeding 100 years even if the initial computational power available is 10^{18} operations per second. This far exceeds the capability of any current computer.

Some encryption systems have systematic weaknesses that allow cryptanalysis without having to try all possible keys (i.e., the structure of the algorithm can be exploited by the cryptanalyst). These systematic weaknesses can greatly reduce the required effort for cryptanalysis and hence the security provided.

Systematic weaknesses can be the result of technical shortcomings in the design of the encryption algorithm, which is the case for the GSM

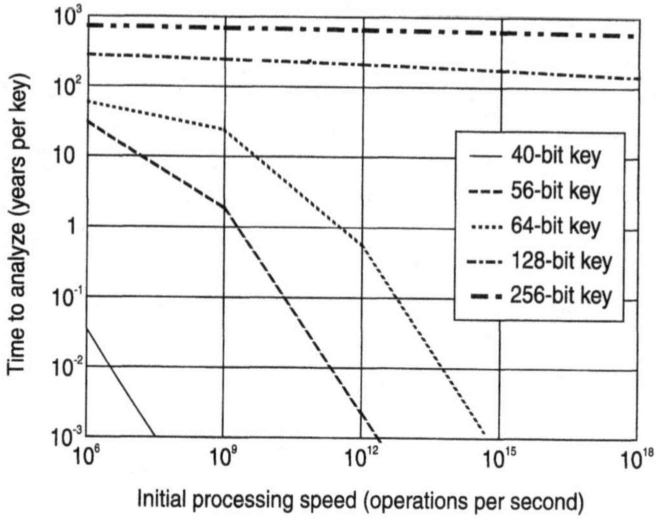

Figure 3.8 Time required for exhaustive cryptanalysis, taking into account Moore's law.

encryption system, known as A5. These weaknesses are discussed further in Section 3.4.7. Systematic weaknesses can also be introduced by procedures. The World War II German Enigma system had two such weaknesses that were crucial in breaking it: The first group of three letters in the plaintext was used to key the deciphering Enigma machine. These three letters were encrypted twice, and the choice of these letters was often predictable. In addition, it was common for some stations to transmit messages beginning with phrases such as "Heil Hitler," providing corresponding plaintext and ciphertext for the cryptanalyst.

3.3.2 Spread-Spectrum Communications

Spread-spectrum techniques are used in communications to provide multiple access, resistance to jamming and other interference, and LPI [8]. In military communications systems, resistance to jamming and LPI (i.e., EP) and multiple access are the major motivations for the use of spread-spectrum techniques. In commercial systems, multiple access and resistance to accidental interference are the major benefits. These features are all achieved by expanding the bandwidth of a signal so that it is transmitted across a number of channels.

There are several types of spread-spectrum techniques. The major types are *direct-sequence spread spectrum* (DSSS), *frequency hopping* (FH), and *chirping.* Chirping is used to provide EP for radar systems, but is not commonly used in communications systems, and is not considered further here.

3.3.2.1 Direct-Sequence Spread Spectrum (DSSS)

Regardless of the type of modulation used, the bandwidth occupied by a digital signal is proportional to its symbol rate. Figure 3.9 illustrates this point for a pseudorandom bit-stream modulated with FSK. When the bit period is halved (i.e., the bit rate is doubled), the bandwidth of the signal around each of the tones is doubled.

A bit-stream can therefore be spread over a larger frequency band by increasing its rate. One way that this can be achieved is to modulate the stream with a pseudonoise (PN) stream at a higher rate. For clarity, the term "bit" is used here to refer to the data unit of the original stream, and the term "chip" is used to refer to the individual data units of the PN stream. The number of chips per bit is often referred to as the *spreading gain* or *processing gain.* The spreading gain is also the factor by which the channel bandwidth is increased.

In a simple example, each 0-valued bit of the stream 01001110 can be replaced with the three chips 010 and each 1 with 101, giving a new chip stream 010101010010101101101010. This stream has three times as many chips as the original stream has bits.

Decoding the DSSS stream requires replacing each group of three chips with either a 0 or a 1 to return the original bit stream. For security, a military DSSS system would usually modify the spreading sequence used from one bit to the next to avoid transmitting a short, easily decoded, repeating pattern.

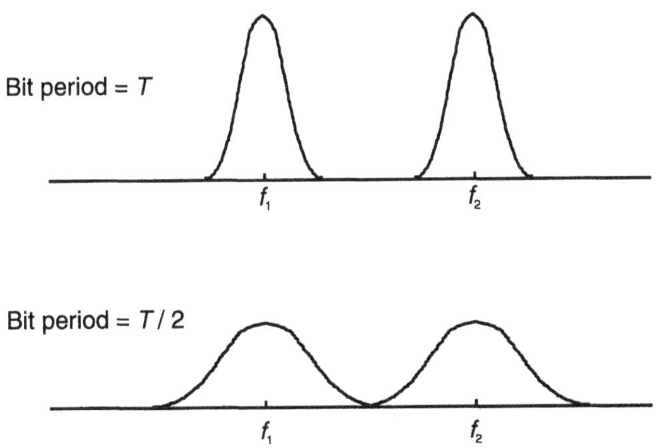

Figure 3.9 Changing bandwidth with symbol time.

Where errors have been introduced into the chip stream, a voting procedure that maps each group of three chips to the closest-valued bit is used.

000 → 0	100 → 1
001 → 1	101 → 1
010 → 0	110 → 0
011 → 0	111 → 1

This process can also be illustrated directly in terms of the bandwidth of the signal. The transmitted spectrum is spread as shown in Figure 3.10 by modulating the baseband signal with the digital code sequence produced by the PN code generator. The DSSS receiver uses the same PN sequence to convert the wideband spread signal back to its original form.

DSSS can be implemented in two ways. In *wideband spread spectrum,* a transmission with a fixed information rate is spread over a number of channels, resulting in an increase in the data rate. In in-band spread spectrum, a single channel is used at a fixed data rate, resulting in a decrease in information rate. This form of spread spectrum permits only very small data rates to be transmitted.

In order to carry out the despreading operation, a DSSS receiver must know the spreading sequence that has been used to spread the original bit-

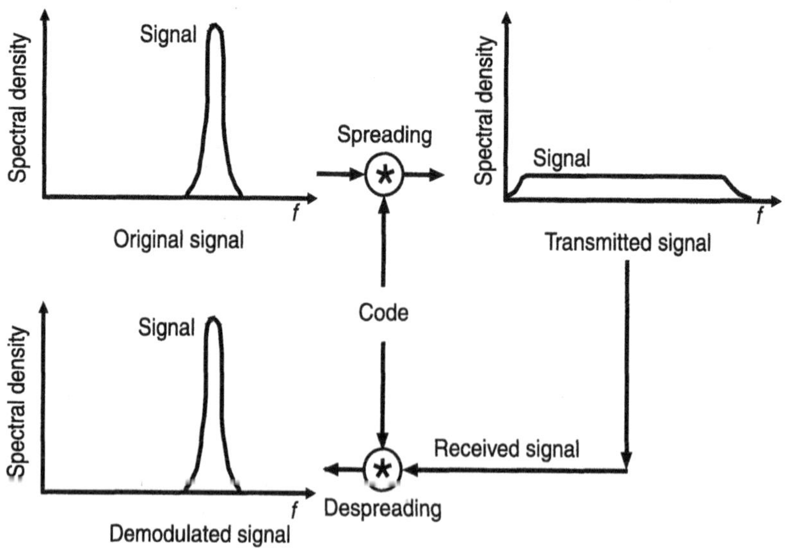

Figure 3.10 DSSS.

stream before transmission. The receiver must also synchronize itself with the transmitter. This may involve knowing (at least implicitly) how many bits have been sent in the current transmission and the location in the PN sequence at which the transmission started. This can be achieved by transmitting a synchronization preamble at the beginning of each transmission, and possibly inserting resynchronization information at regular intervals during the transmission; or using an accurate time reference known to both transmitter and receiver.

Interference from other DSSS signals is covered in this section. Interference caused by single DSSS transmission to other types of signals will have similar effects to a small increase in channel noise, leading to a reduction in the signal-to-noise ratio at the input to the receiver. Where multiple DSSS transmissions take place in a particular part of the spectrum, this noise level will rise significantly and may preclude operation of other systems in the frequency band used.

Resistance to Jamming and Interference

When a spot jammer, or single-channel transmitter, introduces the interference spike shown in Figure 3.11, the receiver despreads the signal but, by the same process, spreads the interference. This means that the receiver can remove most of the interfering signal in the demodulation process. As a

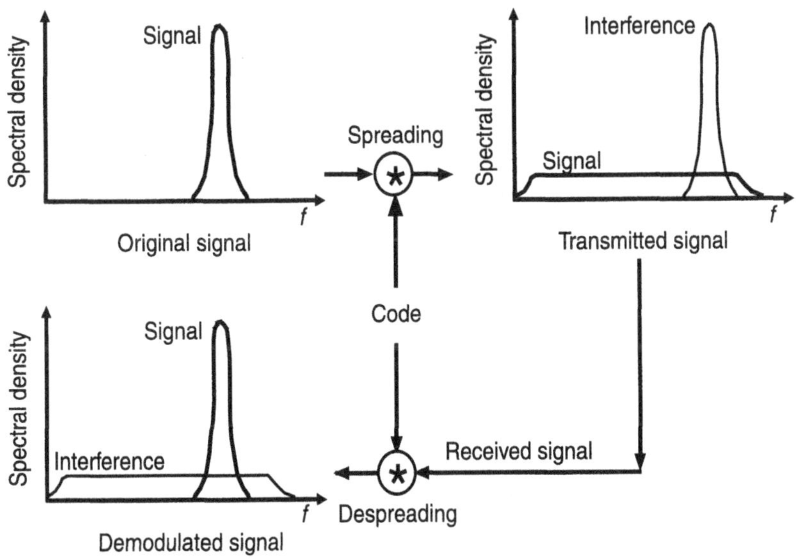

Figure 3.11 DSSS in the presence of narrowband interference.

result, a spot jammer has significantly less impact on the spread signal than on a conventional signal. Additionally, the DSSS signal can occupy the same bandwidth as a number of single channel radios without significant interference.

Figure 3.12 shows that spread-spectrum transmission also provides protection against wideband interference, such as that caused by a barrage jammer or by other spread-spectrum transmissions that use different spreading sequences. Again, any signal not originally spread by the transmitter will not be despread at the receiver, but will be spread to reduce its impact on the wanted signal.

The observations in Figure 3.11 and Figure 3.12 give rise to the concept of *spreading gain.* A signal with original bandwidth f_0 that is spread using DSSS to a bandwidth of f_s is said to have a spreading gain g_s:

$$g_s = \frac{f_s}{f_0}$$

The spreading gain is a measure of the increased tolerance of a spread-spectrum signal to interference compared to a conventional signal. A signal spread using the DSSS transmission will provide the same bit error rate as

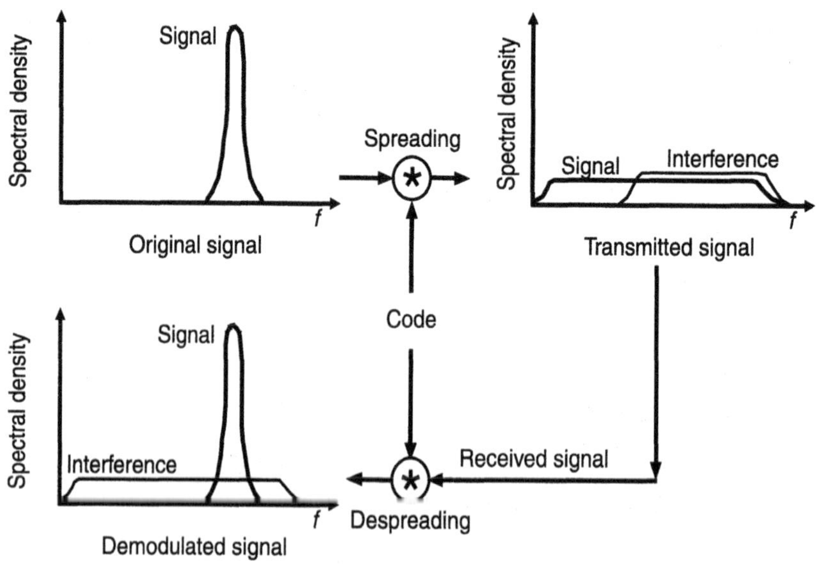

Figure 3.12 DSSS in the presence of broadband interference.

a conventional signal with the in-band interference power higher by a factor of g_s.

For a jammer, this means that where DSSS is used to protect a single signal, a jammer must deliver g_s times as much in-band power to the receiver as would be required to achieve the same effect on a conventional transmission. As the number of DSSS transmitters sharing a part of the spectrum rises, the total jamming power required to achieve a particular effect on each of the DSSS transmissions individually does not rise. As the number of transmitters rises to g_s, the amount of jamming power required to achieve an effect is the same as would be required if the same signals had been multiplexed using FDMA. The gain from the use of DSSS is that multiplexing using FDMA provides the jammer with the option of jamming a subset of the signals and of concentrating limited jamming power on these signals; the use of DSSS aims to reduce the jammer's flexibility by offering only the options of jamming all of the transmissions or none.

DSSS may also be combined with other EP techniques, including FH; the use of error-correcting codes to reduce the impact of residual errors after despreading; and the use of narrowband excision to remove narrowband interfering signals such as those that would be introduced by spot jamming before despreading, leading to an improved signal-to-noise ratio in the receiver.

If a DSSS system uses a preamble to provide synchronization, this preamble may be more susceptible to jamming than other parts of the transmission. If the preamble is successfully jammed, none of the rest of the transmission will be successful. One means of avoiding this vulnerability is to base synchronization on a common, accurate time reference.

Provision of Multiple Access

DSSS can be used to provide multiple access in a radio channel by allocating a different spreading sequence to each transmitter. This system of *code division multiple access* (CDMA) is commonly used in military communications systems as well as in commercial systems such as cellular mobile systems and satellite communications.

Expanding on the simple example used above, three channels may be allocated the (orthogonal) spreading sequences:

Transmitter 1: 0 → 001, 1 → 110
Transmitter 2: 0 → 011, 1 → 100
Transmitter 3: 0 → 111, 1 → 000

These spreading sequences can then be used to convert each of the following four bit streams into a chip stream:

Transmitter 1: 0011 → 001001110110
Transmitter 2: 1010 → 100011100011
Transmitter 3: 1101 → 000000111000

If these three sequences are transmitted onto the same channel at the same power, and the three transmissions arrive synchronized at the receiver, the received chip stream is found to be 000001110010 by selecting the value of a chip as 0 if two or more of the individual channel chips are 0 and 1 otherwise. In practical systems, this voting process occurs when the three modulated signals are combined in the receiver's antenna. Even though the chip timing of the transmissions is synchronized, the carriers of these modulated signals may not be synchronized, causing the received power level to be lowered by destructive interference between the received signals.

The decoding of one channel can be carried out by taking the XOR of the received chip stream with the chip sequence associated with the value 0 for that channel, and applying the same voting procedure as described previously.

For the example above:

Receiver 1: 000001110010 → 001000110011 → 0011
Receiver 2: 000001110010 → 011010101001 → 1010
Receiver 3: 000001110010 → 111110001101 → 1101

In this example, a spreading factor of three has allowed three channels to be multiplexed. The maximum number of bit streams that can be multiplexed onto a single channel, error-free, is equal to the spreading factor.

In many practical situations, the transmissions do not arrive at the receiver synchronized at the chip level. In this case when one channel is to be decoded, the combination of the other interfering signals behaves like noise. The performance obtained depends on the spreading gain and the number of channels being multiplexed; in this case, the high levels of error protection are usually used to reduce the error rate of the output bit stream.

Where DSSS is used to provide multiple access for a number of transmitters all operating at the same data rate, the best performance is obtained when the received power from all DSSS transmitters is equal, and the performance is significantly degraded when this cannot be achieved. In the previous example, if one stream were operating at a power significantly higher than the other two, it would have a greater weight in determining the value of each chip on the channel.

The proportion of the total channel capacity allocated to a stream can be controlled by adjusting its spreading factor and transmit power.

Where DSSS is used as a multiple access technique, a jammer still needs to deliver g_s times as much power to a receiver as would be required for a conventional signal. However, g_s is also the maximum number of signals that can be orthogonally multiplexed. When CDMA is used to full capacity, the difficulty of jamming all these signals is the same as jamming the same set of signals using conventional modulation. The difference is that the jammer does not have the option of concentrating power into one (high-priority) signal; it is forced to jam all signals or none. This effectively prevents a jammer from obtaining the benefits of spot jamming.

Impact of Overloading

Unlike FDMA and TDMA, CDMA allows (at least theoretically) over-allocation of channel capacity. CDMA can support an essentially arbitrary number of streams, so long as each stream is allocated a different spreading sequence. There is, however, a strict limit on the number of streams that can be carried without mutual interference. The spreading gain g_s, which is equal to the maximum number of orthogonal PN sequences, provides an upper bound on this number of streams. In the absence of external interference or jamming, this limit is equal to g_s. In the presence of external interference, the number of streams that can be carried without error is less than g_s.

Frequency Management

Spectrum management for DSSS transmissions involves managing both frequencies and spreading sequences. Interfering signals in a DSSS system have a similar impact to an increase in channel noise. This means that when the system approaches its capacity, the quality is gracefully degraded. Frequency management in DSSS systems usually takes advantage of this graceful degradation, implementing statistical rather than deterministic separation between signals. This has the potential to greatly simplify the frequency management process, minimizing the need for procedures such as the analysis of harmonics and intermodulation products.

Near-Far Effect

A requirement for efficient implementation of CDMA (i.e., the use of DSSS as a multiple access technique) is that each transmitter adjusts its transmitted power so that all receivers receive the same power from each transmitter. This requirement for power balancing has two key implications. First, such power balancing is not possible where there are multiple transmitters and multiple receivers. This makes CDMA unsuitable as a multiple access technique for net-oriented communications, such as CNR. This effect is known as the *near-far effect.* Second, successful implementation requires continuous monitoring of received power levels by receivers. The measured levels are sent to transmitters on a signaling channel.

Power balancing is used in CMDA cellular telephone systems to overcome the near-far effect. This is only possible because all communications are either to or from a base station; there is no direct mobile-mobile communications.

The power balancing problem is illustrated in Figure 3.13. If stations 1 and 3 set their transmit power so that same power is received at station 2, then station 1 cannot adjust its power to achieve the same received power at station 3 as station 2. The only solutions to this problem are to accept a loss in efficiency or to structure the system so that all CDMA operates with a single transmitter or a single receiver. This can be achieved in systems that require all communications to pass through a base station, such as in mobile telephone networks. However, it cannot be achieved satisfactorily in CNR nets.

Protection Against ES

As shown above, the use of DSSS spreads the power of a transmitted signal over a large band. This spreading increases the difficulty of detecting the signal. Typically, a DSSS signal cannot be detected by a narrowband search receiver. Detection by a wideband search receiver can sometimes be achieved by measurement of power levels across a wide band of frequencies.

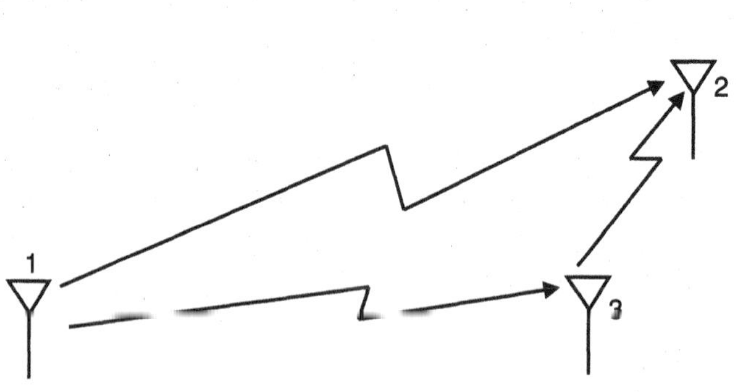

Figure 3.13 The power-balancing problem to overcome the near-far effect.

A DSSS transmitter provides two types of protection against DF. First, a DSSS signal must be detected before direction-finding techniques can be applied. Second, a DSSS signal must be separated from other in-band interfering signals (including both DSSS and conventional signals) before DF is possible. DF techniques able to operate on multiple cochannel signals, such as Doppler DF, may be capable of providing this separation internally. For widely spaced transmitters, the removal of these interfering signals may be achieved by the use of directional antennas. For conventional interfering signals, the use of narrowband excision may also be an option.

Signal interception is hindered both by the difficulty of detection and by the fact that received chips need to be despread before decoding. This can only be achieved if the spreading PN sequence is known or can be inferred from the received data. Military systems are usually designed with very long PN sequences to maximize the difficulty of intercept. Commercial systems, however, tend to use shorter spreading sequences, which may be known in advance or easily inferable from the received signal.

In conventional half-duplex communications systems, a user community (or net) consists of those stations that share a particular channel. When DSSS is introduced, transmissions on a net are no longer confined to a single frequency, removing this means of identifying user communities.

3.3.2.2 Frequency Hopping

Frequency hopping (FH) is a form of spread-spectrum communications where the transmitter periodically changes the frequency of transmission, as illustrated in Figure 3.14. By knowing the hopping sequence, a receiver follows the changes in frequency and is able to receive the transmission. A nonhopping receiver is unable to receive data transmitted by a hopping transmitter. The effectiveness of FH relies on having a large set of frequencies in the hop set and on the pattern of frequency changes appearing to be random.

In order to provide resistance to jamming, it is necessary that a pseudorandom hop sequence be used. This requires that a receiver synchronize itself to the transmitter, which can be achieved by transmitting a preamble at the beginning of each transmission and possibly the transmission of further synchronization information at regular intervals during the transmission, or using an accurate time reference known to both transmitter and receiver.

The term *hop rate* refers to the number of times per second that an FH transmitter changes frequency. *Dwell time* is the reciprocal of hop rate and is the time interval in which data is transmitted between consecutive changes of frequency.

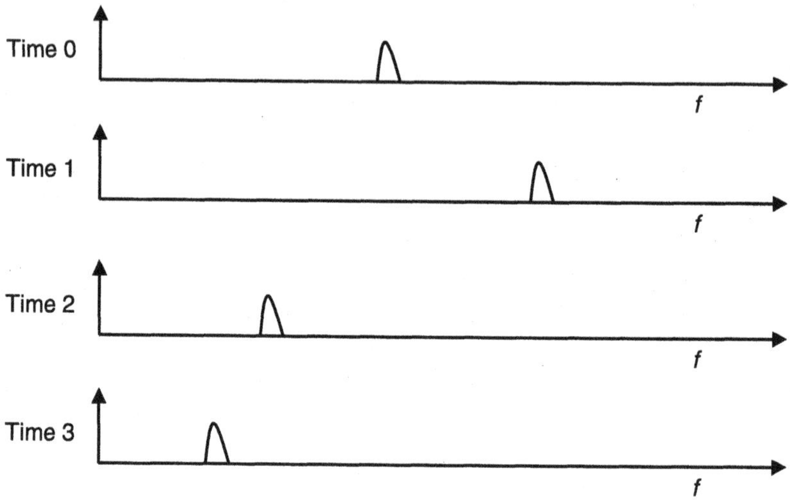

Figure 3.14 Effect of FH.

The terms *slow* and *fast* hopping are sometimes used. Fast FH refers either to a hop rate that is higher than the bit rate or to a hop rate significantly faster than 100 hops per second for VHF and UHF transmissions, and 10 hops per second for HF transmissions. Slow FH refers to slower hop rates.

In most FH systems, there is a *guard interval* during which no information is transmitted while the transmitter changes frequency. This is required because the transmitter's power amplifier cannot instantaneously change frequency. In order to avoid sweeping its power across the band and causing widespread interference, the transmitter stops transmitting during the period of the frequency change.

One of the advantages of FH is that it is possible to choose the hop set from an arbitrary list of channels. Unlike DSSS, there is usually no requirement for these channels to be adjacent.

FH may also be combined with other EP techniques, including DSSS and the use of error-correcting codes and interleaving to reduce the impact of errors caused by clashes with other transmissions.

Resistance to Jamming and Interference

In order to jam a hopping transmission effectively, a jammer must either generate power across a high proportion of the frequencies used (which implies a large total power) or be able to follow the frequency changes. A fixed-frequency spot jammer will have minimal impact on an FH transmission.

The number of channels in the hop set is analogous to the spreading gain g_s of a DSSS system, and is a measure of the increased difficulty of jamming using jammers that are unable to follow the hopping. The gains that can be achieved with FH may be much larger than can be achieved with DSSS. It is common, for example, for FH VHF CNR to be able to hop across the whole of the 30–88 MHz band.

Effective jamming of a hopping transmission requires that the jamming signal arrive at receiver before end of transmission in one hop. Unless the frequency in use is known in advance, the following sources of delay occur between the beginning of transmission and the jamming signal arriving at the receiver:

- The propagation delay difference between the path from transmitter-jammer-receiver and the direct path from transmitter to receiver:

$$t_{propagation} = \frac{distance_{Tx-J} + distance_{J-Rx} - distance_{Tx-Rx}}{3 \times 10^8}$$

- The processing delay at the ES receiver, which is greater than the reciprocal of the receiver bandwidth;
- The jammer's power-up delay.

Successful jamming requires that the sum of these delays be a small enough proportion of the dwell time that a sufficient portion of the transmitted signal is jammed.

Table 3.1 shows the maximum distances over which it is possible to jam a single hop as part of an FH transmission, based on the assumptions

Table 3.1
Maximum Distances (km) over Which a Frequency-Hopping Transmission Can Be Jammed

Hop Rate (hops per second)	Receiver Bandwidth (kHz)			
	25	50	100	Infinite
100	992	996	998	1,000
500	192	196	198	200
1,000	92	96	98	100
10,000	2	6	8	10

that the receiver is close to the transmitter, the length of the guard interval is zero, all delay is due to propagation and the detection time of the receiver (i.e., the jammer power-up delay is zero), and the jamming is effective if one-third of the hop time is jammed.

From Table 3.1, it is clear that even for a rate of 100 hops per second (a dwell time of 10 ms), the use of satellite jammers is unlikely to be possible. Standoff airborne or ground-based platforms, however, are likely to be within this range limit. At 10,000 hops per second, standoff airborne platforms (e.g., high altitude UAVs) are also infeasible. Jammers, such as UAJ, capable of close-in deployment are the only systems likely to be effective against such high-speed hoppers.

Provision of Multiple Access

FH can also be used as a means of multiple access by allowing several nets to share frequencies in their hop sets. As long as these frequencies are used in a unique sequence by each net, a statistical separation between nets is provided.

Impact of Overloading

As the number of FH transmitters sharing common frequencies increases, the proportion of the time during which two or more transmitters clash on a single frequency increases. Under these circumstances, it is possible that neither signal will be correctly received.

Voice For voice, loss of data due to clashes causes glitches that may be annoying. The comprehensibility of voice, however, tends to degrade gracefully. This is because even though the voice is digitally encoded and compressed, the human auditory system is quite forgiving of errors introduced in transmission. Voice tends to become unusable only when approximately one-third or more of all data is lost due to interference from other hopping or nonhopping signals. Since it is difficult for a jammer to maintain power levels over one-third of the hopping band, FH provides robust voice performance in a hostile electromagnetic environment.

Data Loss of data due to clashes in the use of a frequency is a particular problem for data communications, especially on slow hoppers. Unlike voice services, quality-of-service can be significantly degraded for data services with interference due to clashes or jamming in only a small proportion of hops. Some systems, such as those used for situational awareness, overcome this problem by providing regular repetition of information. Protection can also be provided using a combination of interleaving and forward-error correction.

Frequency Management

Frequency management for FH requires the allocation of a hop set to each hopping transmitter. One or more frequencies in this set can be allocated to more than one net, as long as each net is allocated a different hop sequence. In other words, separation between systems can be provided by a choice of different frequencies in the hop sets or by the use of the same frequencies in a unique sequence for each net. There is therefore a reduced need for frequency deconfliction between nets, including processes such as harmonic analysis that are an important part of conventional frequency management. There is also no extra management overhead if the capacity allocated to the nets varies.

Interference with nonhopping systems is minimized by excluding frequencies used by these systems from the hop set. The acceptability of these clashes depends on the particular service. Some channels, such as those used for emergency communications, are usually excluded completely from hop sets.

Interference with other hopping systems is minimized by choosing hop sets with no frequencies in common. However, this will only be possible on very rare occasions. More commonly, interference is managed by ensuring that the proportion of hops for which two or more hoppers in the same area will share a common frequency does not exceed a quality-of-service threshold. This issue is discussed further in the next section.

Near-Far Effect

Unlike systems using either DSSS or TDMA, FH communications are not subject to the near-far effect. This is because no attempt is made to synchronize FH transmitters on different nets, and multiple access using FH is based on FDMA, which does not require synchronization.

Protection Against ES

FH transmissions are relatively difficult to detect using narrowband search receivers. This is because detection only occurs when the frequency of the hopping transmission coincides with the frequency of the scanning receiver. A wideband receiver is more suitable for the detection of FH transmissions because it can effectively monitor a number of channels simultaneously.

DF of an FH transmitter typically requires detection and DF to occur within the dwell time. This suggests that a high level of integration between search and DF is required.

Intercept of an FH transmission can be achieved by using a multichannel receiver with a directional antenna. Such systems are most successful where there is only a single FH transmitter active in a band at any time or the frequency hoppers are widely dispersed. Consequently, the LPI properties of FH are enhanced by the use of multiple FH nets that share at least some frequencies in their hop sets.

In conventional single-frequency, half-duplex communications systems, a user community (or net) consists of those stations that share a particular channel. When FH is introduced, transmissions on a net are no longer confined to a single frequency, removing this means of identifying user communities. FH systems, however, that transmit a preamble either on a fixed frequency or in some other identifiable way, may lose this protection.

3.3.2.3 Comparison of Spread-Spectrum Techniques

The advantages of DSSS are that it hinders adversary ES, making transmissions hard to detect above the noise floor; hinders adversary EA; provides graceful degradation in the presence of jamming and other types of interference; provides multiple access, known as code division multiple access (CDMA); and allows mutual interference to be managed more easily than in FH.

The disadvantages of DSSS are that the near-far effect makes power management of multiple access in multinet environment difficult; mutual interference contributes to total jamming power; DSSS provides less efficient multiple access than TDMA or FDMA; and spreading usually occurs over contiguous band (i.e., most practical systems do not allow spreading over an arbitrary collection of channels). In-band spread spectrum has the additional disadvantage of providing only very low data rates.

The advantages of FH are that it hinders adversary ES/EA; the hopping frequencies can be chosen from arbitrary set; the channels used do not need to be contiguous; and it can be used as a multiple-access technique.

The disadvantages of FH are that hopping nets will usually share some or all of the frequencies in their hop set, resulting in mutual interference between hopping nets interference with nonhopping nets whose frequencies are included in the hop set of a hopping net; and ES techniques, based on using multichannel receivers and directional antennas, and EA techniques, based on follower-jammers, exist that can overcome the EP provided by slow-hopping transmitters, especially when used in isolation.

3.3.3 Error-Protection Coding

Digital signals passing through transmission channels are subject to errors introduced in the transmission process. While all channels introduce errors, some channels, especially radio channels subject to jamming, will introduce very high error rates. The types of impairments introduced by an imperfect transmission channel include additive noise and channel perturbations. Additive noise may take the form of Gaussian noise with stationary statistics, impulsive noise that is not always stationary or easy to characterize, or jamming. Channel perturbations may occur due to fading in radio channels, synchronization slip in digital channels, and breaks in the transmission of diverse origin.

The likely effects of these channel impairments on a digital signal are *uniformly random errors*—errors occurring individually and independently, with approximately uniform probability density, primarily due to noise (often just called *random noise*); *burst errors*—errors grouped in clusters, mainly the result of a combination of noise and channel perturbations; and *erasures*—irregular intervals when it is known that no reliable signal can be detected, because of severe channel perturbation.

Channel coding is used to correct errors caused by these channel impairments through the introduction of controlled redundancy to enable messages corrupted in transmission to be corrected before further processing [9]. With this controlled redundancy, only a subset of all possible transmitted messages (bit sequences) contains valid messages. This subset is called a code, and the valid messages are called *code words* or *code vectors.* A good code is one in which code words are so separated that the likelihood of errors corrupting one into another is kept small.

Error detection is simplified to answering this question: Is the received message a code word or not? If it is a code word, one assumes that no errors have occurred. The probability of an undetected error getting through is then the probability of sufficient errors occurring to transform the real transmitted code word into another, apparently correct, but in reality a false one.

If an error is detected, it can be corrected in principle by automatic repeat request (ARQ) or forward error correction (FEC).

- *Automatic repeat request (ARQ).* In ARQ, the recipient rejects the received message as erroneous and requests a repeat transmission. However, if propagation delays due to distance are large, the technique may become so inefficient as to be useless. There are also many cases where retransmission is impossible, such as extracting information from a damaged archive.
- *Forward error correction (FEC).* In FEC, the recipient corrects the errors by finding the valid code word "nearest" to the received message, on the assumption that the nearest is the most likely because few corrupting errors are more likely than many.

There are two types of FEC: *block coding* and *convolutional coding.* In block coding, source data is partitioned into blocks of *k* bits, converted by the encoder into blocks of *n* (>*k*) bits with enough checks to enable the decoder to correct errors of the more probable kinds. Error-correcting codes have more redundancy than error-detecting codes, and the decoding algorithms are much more complex. The most common types of block codes [9] are cyclic redundancy check (CRC) codes, which provide only error detection; Golay codes; Bose-Chadhuri-Hocquenghem (BCH) codes; and Reed-Solomon (RS) codes. For a convolutional code, the encoder operates not on disjoint blocks, but on a running block of bits held in a shift register, generating a sequence of higher rate. This procedure is normally used for FEC, but the correcting capabilities are not so clear-cut as with block codes. Probabilistic decoding, approximating the maximum likelihood, is generally used.

Block codes are used when information is naturally structured in blocks, when channel capacity is relatively low and we do not want to waste it further with unnecessarily low code rates, and when quick efficient decoding is required because of the limited processing time available.

When long streams of relatively unstructured data are transmitted on high-capacity channels (e.g., satellite 10 Mbps channels) and when the complexity of the decoder represents a relatively small proportion of the total cost of the receiving equipment (e.g., a satellite receiver), convolutional codes can offer the best error-correcting solutions.

Interleaving

Error-correcting codes can be used to detect and correct random bit errors. The codes are effective as long as the number of errors close together remains small. In many types of channels, especially radio channels, however, the channel errors occur in bursts of many errors followed by long periods with almost no errors.

The problem of bursty channel errors can be overcome by interleaving the transmitted data. This is achieved by rearranging the coded data at the transmitter in a predefined pseudorandom order. This means that a burst of errors will be randomized at the receiver when the bits are placed back in their original order.

Concatenated Codes

Concatenated codes use two levels of coding—an inner code and an outer code—to achieve the desired error performance. As illustrated in Figure

3.15, the inner code is configured to correct most channel errors; the outer code reduces the probability of error to an acceptable level.

One of the most popular systems uses a convolutional inner code and a Reed-Solomon outer code. The Reed-Solomon coder is chosen because it can operate on symbols that consist of a number of bits. Like other FEC, it operates best on isolated symbol errors. Because the symbols may consist of a number of bits, the Reed-Solomon coder is quite effective at correcting bursts of bit errors.

3.3.4 Burst Transmission

Many communications systems transmit continuously, making them easy targets for adversary ES, especially DF, and EA. Even systems that only transmit when they have data to send have traditionally used voice transmission, resulting in lengthy transmissions that once again present good targets to adversary ES and EA.

The term *burst transmission* is used to refer to systems that provide protection against ES and EA by transmitting for short periods only when they have data to pass. At its simplest, burst transmission may be used to describe the operation of a data-enabled CNR system. More sophisticated covert communications systems based on burst transmission use the combination of high data rate transmission and low transmission time as a means of frequency spreading. This may be used in conjunction with other EP techniques, such as error protection, spread spectrum, and encryption.

The effectiveness of the burst transmission as a form of EP is controlled by the length of the burst and the predictability of its transmission. The length of the burst depends on the amount of data to be transmitted, overheads such as error protection, the transmit data rate, and the length of any preambles required for synchronization. Parameters associated with the predictability of the transmission include the time of start, duration, carrier frequency, code (if DSSS is used), and bandwidth of the transmission.

The effectiveness of burst transmission can be increased by delays in handing over targets within the EW system. For example, in a system with separate search and DF systems, a burst transmission may be over before the DF has even received the target information. Because of the short transmission

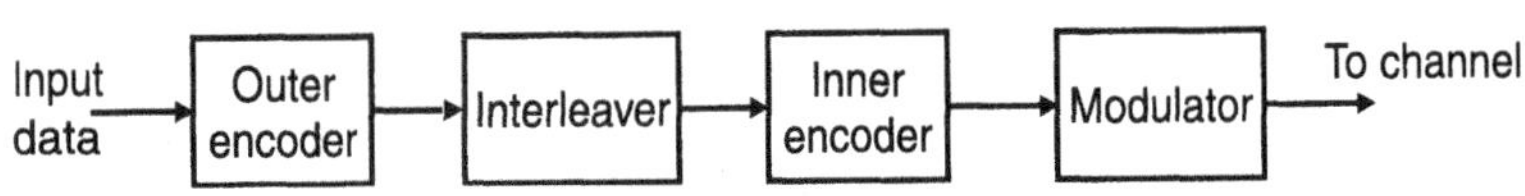

Figure 3.15 Block diagram of a concatenated coder.

times, similar constraints to those discussed for FH apply to the maximum range over which a jammer can be effective against burst transmissions.

3.3.5 Narrowband Excision

A jammer may sometimes transmit in only a portion of the band being used for communications. This is most likely to happen in higher capacity channels, such as those used for radio relay or DSSS signals. This is illustrated in Figure 3.16. Such jamming can be effective against conventional receivers. Even for a DSSS receiver, the power of the jamming signal is turned into noise in the demodulation process.

In a receiver employing narrowband excision, narrowband jamming is detected and removed from the signal before it is passed to the demodulator as illustrated in Figure 3.17. While this also removes any information in the excised band, the overall effect is beneficial because the jammer power is prevented from having any impact on the demodulation process. Because the loss of some portion of the received signal is a necessary side effect of narrowband excision, some means for correcting the resulting errors are usually required.

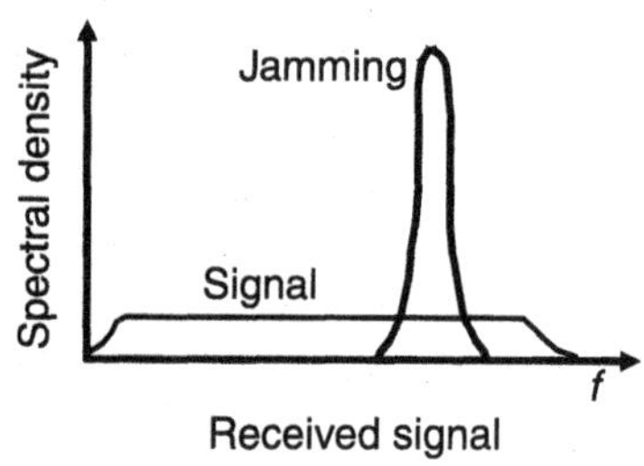

Figure 3.16 Narrowband jamming of a wideband signal.

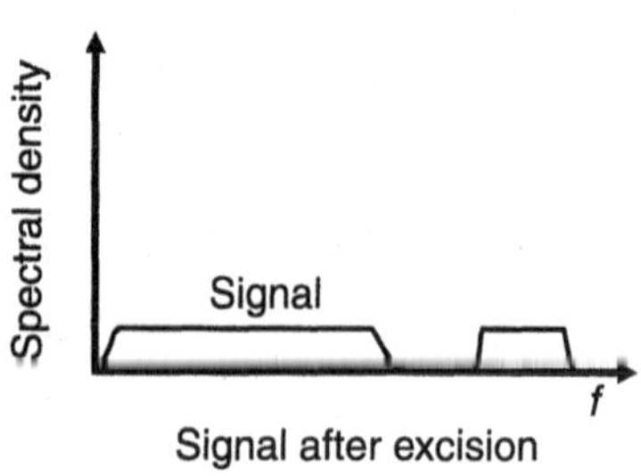

Figure 3.17 Signal after narrowband excision.

3.3.6 Diversity

Diversity is a means of protecting against jamming and interference by transmitting data over two or more channels [10]. There are three types of diversity commonly used in communications systems: space, frequency, and time.

- *Space diversity.* The impact of narrowband fading in a communications channel can change significantly when the location of the receiver's antenna is moved by more than half the wavelength of the signal. In space diversity, two or more receive antennas are placed more than half a wavelength apart, effectively providing two channels between transmitter and receiver. This type of diversity is commonly used in microwave transmission systems.
- *Frequency diversity.* By transmitting data on two or more different frequencies, the impact of fading or jamming and interference on one of the channels can be overcome. This type of diversity is used in HF sky-wave communications, including TADIL-A (Link-11) [11].
- *Time diversity.* Data can be transmitted twice on the same channel, protecting against short-term jamming and interference.

3.4 Use of EP Techniques in Communications Systems

Many military and commercial communications systems employ one or more EP techniques. In this section, a selection of these systems is reviewed, highlighting the differences between the military and commercial use.

3.4.1 CNR

VHF CNR typically operates in the frequency range of 30 to 88 MHz. Modern CNR systems provide both secure voice and secure data capabilities. Most systems, such as the US SINCGARS, also provide an in-built FH capability. In some radios, the FH is provided by an appliqué. Free channel search is also found in some systems.

CNR does not usually employ DSSS because the use of DSSS would require that channels much larger than the conventional 25 kHz be allocated to each net, reducing the overall capacity in the 30–88 MHz band; and the near-far effect prevents the effective use of CDMA in CNR to improve the efficiency of use of the electromagnetic spectrum.

Encryption is commonly provided either using in-built systems or an appliqué. Traditionally, such encryption has been used at battalion and higher levels, leaving lower-level transmissions open to intercept. Increasingly, encryption is being pushed down to the lowest levels.

Because the encryption used in CNR is message-based, the synchronizing preambles of these transmissions are particularly vulnerable to jamming. Adversary EA may deliberately jam preambles to force nets to operate in plaintext, allowing their transmissions to be intercepted.

Error protection is not commonly used for voice transmissions in CNR. The digital voice coding algorithms used tend to provide very high levels of robustness against transmission errors, operating satisfactorily with bit error rates as high as 10%. However, some form of error protection may be provided for data.

3.4.2 Military Radio Relay

Military radio relay systems operate in the VHF and UHF bands above 200 MHz. A network is formed from a number of point-to-point links that interconnect nodes that perform switching. Traditionally, these systems have provided circuit-switched voice and data services.

Because they are based on point-to-point links, radio relay systems almost always use directional antennas. The gain of these antennas is typically around 10 dB. These directional antennas provide a high level of protection against adversary EA and ES.

Encryption in radio-relay networks is based on bulk encryption of the point-to-point links, with switching occurring on plaintext data, which protects against vulnerabilities associated with the restarting of cryptographic algorithms in message-based encryption. FH and DSSS transmission are also sometimes employed in these systems to provide additional protection. Additionally, many radio-relay systems, especially those that provide for the carriage of data as well as voice, provide some form of error protection, which is commonly based on a half-rate convolutional code.

3.4.3 TADIL J (Link 16)

TADIL-J is a secure, high-capacity, jam-resistant, nodeless data link that uses the Joint Tactical Information Distribution System (JTIDS) transmission characteristics and the protocols, conventions, and fixed-length message formats defined by the JTIDS Technical Interface Design Plan (TIDP) [12]. TADIL-J operates in the UHF band in the frequency range of 960 to 1,215 MHz, and therefore provides line-of-sight operation. Operation beyond line-of-sight can be achieved by means of a relay, which may be an airborne or satellite-mounted system.

TADIL-J operation is based on all-informed nets. Multiple access between nets is provided by a combination of FH, FDMA, and CDMA. There are 51 channels that are supported. Multiple access within a net is provided by TDMA. The TDMA structure is shown in Figure 3.18. *Time slots* of 7.8125 ms are allocated to stations on the net. Approximately 1,536 time slots make up a *time frame,* and 64 time frames form an *epoch.* Each station on the net is allocated at least one time slot per epoch.

Table 3.2 shows the TADIL-J maximum range and data rates, which depend on the operating mode.

EP in TADIL-J is provided by a combination of FH with an instantaneous hop rate of 77,000 hops per second over 51 frequencies; DSSS with a spreading gain of 6.4 and a chip rate of 5 MHz; repeated transmission with data optionally transmitted twice in successive hops; and forward-error detection and correction, using a (31, 15) Reed-Solomon code.

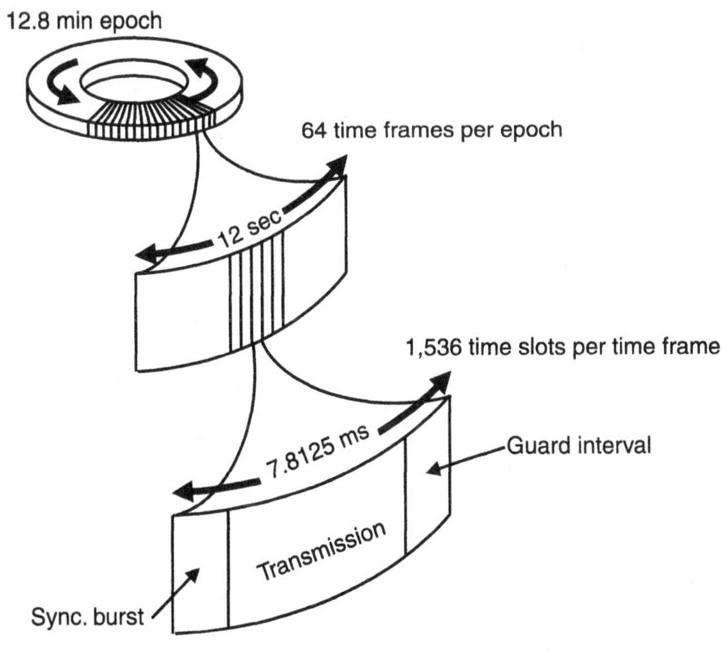

Figure 3.18 TADIL-J time-slot structure.

Table 3.2
TADIL-J Operating Modes

Mode	Guard Interval (ms)	Guard Interval Range Limit (nm)	Throughput After Error Correction (Kbps)	Hops per Second
Standard full slot	4.4585	700	30	33,000
Packed-2 full slot	4.4585	700	59	33,000
Packed-4 full slot	2.0405	300	119	57,000

In each hop, the transmitter is turned on for 6.4 μs, which means that power from a jammer with transmitter-jammer-receiver path length 2 km longer than the transmitter-receiver path length will reach the receiver after the end of the data transmission.

3.4.4 Enhanced Position Locating and Reporting System (EPLRS)

EPLRS is a U.S. situational awareness system that is designed to provide services for position location, navigation, identification, and communications. EPLRS also supports a number of control measures such as boundaries, fire support coordination lines, limits of advance routes, passage lanes, and attack directions.

The EPLRS radio supports a variety of data communications services [13], providing both point-to-point links and an extensive multicast capability, including all-informed nets. Data rates up to 57,600 bps per connection, known as a needline, are possible. Each EPLRS user community has a maximum practical data capacity of between 300 and 450 Kbps, depending on configuration. This capacity is reduced when retransmission is required. Capacity is also reduced when the area over which the user community increases, because of the requirement for larger guard intervals between TDMA slots.

Frequencies in the range of 420 to 450 MHz are used, with the band being segmented into eight channels, each of 3-MHz bandwidth. Multiple access within a net is provided by TDMA technology, in which users transmit information in bursts during predetermined time slots. Multiple access between nets is provided by a combination of FDMA, FH, and CDMA.

Resistance to jamming is provided through DSSS transmission with a spreading gain of approximately five; FH operation among eight channels with a hop rate of 512 Hz; error detection and correction; and network management that facilitates the automatic routing and rerouting of messages in the EPLRS network using any EPLRS radio as a relay of opportunity.

In each hop, the transmitter is turned on for up to 1.1 ms, which means that the transmitter-jammer-receiver path length must be no more than 33 km longer than the transmitter-receiver path length if jamming is to be effective.

The combination of DSSS and FH also provides LPI. Security is provided by an embedded cryptographic system. While the use of DSSS provides a multiple-access capability, the near-far effect will limit its efficiency. The use of DSSS also provides protection against interference from other EPLRS nets. This would otherwise be a significant problem because only eight channels among which the transmitter can hop are available.

3.4.5 Near-Term Digital Radio (NTDR)

NTDR is an experimental system being developed by the U.S. Army under the Force XXI program to explore the limits of near-term technology and to provide a technical baseline for development of a multiband, multimode digital radio system.

NTDR can be viewed as an RF system with an embedded router/gateway such as those found in fixed local area networks. NTDR transports up to 288 Kbps of user information for each cluster of users, backbone channel, or point-to-point connection for the operating frequency range of 225 to 450 MHz with a channel bandwidth of 4 MHz.

EP will be provided by a combination of DSSS, fast FH, and narrowband excision at receivers to eliminate effects of narrowband jamming. Further protection against error will be provided by three-quarter-rate convolutional coding. Once again, the combination of DSSS and FH is designed to minimize the detectability of signals. The use of narrowband excision allows jamming signals from narrowband jammers to be removed before despreading, reducing even further the impact of such signals.

3.4.6 IS-95

IS-95 is an air-interface standard for cellular telephony. It uses a combination of CDMA and FDMA to provide multiple access on both downlinks (forward channels) and uplinks (reverse channels).

Data on the forward channel is grouped into 20-ms frames. This data is convolutionally encoded, repeated if necessary to increase the data rate to 19.2 kilosamples per second (ksps) and interleaved, as illustrated in Figure 3.19. The signal is randomized with a long PN sequence and spread with a Walsh code to produce a 1.2288 megachips per second (Mcps) signal.

Power control information is inserted every 1.25 ms by puncturing. Mobile terminal transmit power is adjusted in 1-dB steps. This high-rate, fine adjustment or transmit power is required to provide power balancing between mobile stations and to maximize the bandwidth efficiency of the system.

Quadrature modulation is performed as shown in Figure 3.20, with in-phase and quadrature components having an orthogonal covering applied

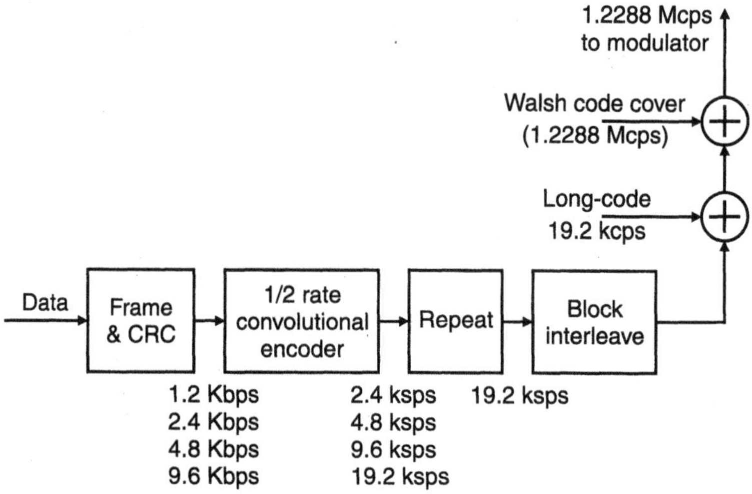

Figure 3.19 Downlink processing for IS-95.

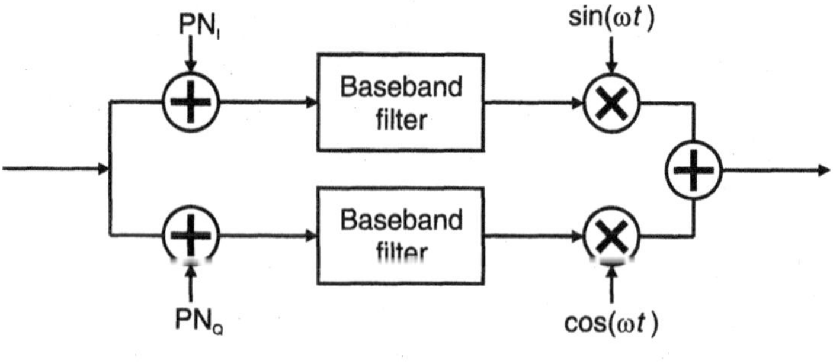

Figure 3.20 Downlink modulation for IS-95.

before modulation. At the receiver, this makes the received components approximately independent. The covering is performed with a short code that is relatively easy for a receiver to acquire.

The use of DSSS in IS-95 is purely intended to provide multiple access. Because the PN sequences are known or easily deducible, IS-95 does not provide an LPI or antijam capability, except against a very unsophisticated attack. The use of high levels of forward-error correction is intended to overcome the noise introduced by interfering signals, both narrowband and wideband.

3.4.7 GSM

The GSM digital cellular telephone system provides multiple access using a combination of FDMA and TDMA. Eight TMDA channels are multiplexed onto a carrier with a channel bandwidth of 200 kHz. Each time slot contains 114 bits of user data [14].

GSM provides an optional FH mode. This mode does not provide an LPI or antijam capability, because the specification for the hop sequence is published in the GSM standards, and the hop sequence can therefore be deduced from the signals transmitted by a base station.

FH in GSM does, however, provide frequency diversity. This diversity is intended to minimize the impact of multipath propagation, which may lead to much higher losses in some channels than in others. Because GSM was not designed for military use, no extra benefit was perceived for LPI or antijam capabilities.

Error protection in GSM takes 240 bit blocks of data, and codes them with a half-rate punctured convolutional code to produce 456 bits that are interleaved across four 114-bit TDMA frames. This interleaving spreads burst errors (that are caused largely by channel fading) over a longer period, reducing the reduce peak bit error rate, and allows the channel coding to correct the now-randomly spaced bit errors.

Encryption in GSM is based on a proprietary, stream-cipher algorithm, known as A5. A5 comes in two variants: A5/1 is used in European systems and A5/2 (which is known to provide a significantly lower level of security) is used in export systems. A5 is a stream cipher whose state is reinitialized at the beginning of every TDMA time slot. Its vulnerabilities [5] include the fact that in most deployed versions of GSM, the 10 least significant bits of the key are set to zero, reducing the effective length of the key to 54 bits; the keystream is frequently reinitialized, permitting attacks based on a known initial state; while the state transition function of A5 is not uniquely invertible, it can be efficiently inverted because the number of possible parent states is small; and cryptanalysis of A5/1 requires approximately 2^{24} operations (with 2^{48} precomputed stored values), while cryptanalysis of A5/2 requires only 2^{16} operations.

Endnotes

[1] U.S. doctrine for EP is contained in U.S. Army Field Manual FM 24-33, "Communications Techniques: Electronic Counter-Countermeasures," July 1990.

[2] *Electromagnetic Pulse (EMP) and TEMPEST Protection for Facilities*, Engineering and Design Pamphlet EP 1110-3-2, U.S. Army Corps of Engineers, December 1990.

[3] Sources of information on encryption techniques include:
Denning, D. E., *Cryptography and Data Security*, Reading, MA: Addison Wesley, 1983.
Schneier, B., *Applied Cryptography: Protocols, Algorithms, and Source Code in C*, New York: Wiley, 1994.
Singh, S., *The Code Book: The Science of Secrecy from Ancient Egypt to Quantum Cryptography*, New York: Anchor Books, 1999.
Sinkov, A., *Elementary Cryptanalysis, a Mathematical Approach*, New York: The Mathematical Association of America, 1966.
Stallings, S., *Network and Internetwork Security*, 2nd Edition, Englewood Cliffs, NJ: Prentice Hall, 1995.
Torrieri, D. J., *Principles of Secure Communications*, Norwood, MA: Artech House, 1985.

[4] *SKIPJACK and KEA Algorithm Specifications*, Version 2, National Security Agency, May 1998.

[5] Biryukov, A., A. Shamir, and D. Wagner, "Real Time Cryptanalysis of A5/1 on a PC," *Fast Encryption Software Workshop 2000*, New York, April 2000.

[6] Stallings, S., *Network and Internetwork Security*, 2nd Edition, Englewood Cliffs, NJ: Prentice Hall, 1995.

[7] *SKIPJACK and KEA Algorithm Specifications*, Version 2, National Security Agency, May 1998.

[8] See, for example, Simon, M. K., *Spread Spectrum Communications*, Rockville, MD: Computer Science Press, 1985; or Nicholson, D. L., *Spread Spectrum Signal Design: LPE & AJ Systems*, Rockville, MD: Computer Science Press, 1988.

[9] See, for example, Sklar, B., *Digital Communications*, Englewood Cliffs, NJ: Prentice Hall, 1988; and Proakis, J. G., *Digital Communications*, 2nd Edition, New York: McGraw-Hill, 1989.

[10] See, for example, Gibilisco, S., *Handbook of Radio and Wireless Technology*, New York: McGraw Hill, 1999, pp. 252–253.
ITU-R, Recommendation F.106-2, 1999.

[11] MIL-STD-188-203-1A "Interoperability and Performance Standards for Tactical Digital Information Link, (TADIL) A," January 1988.

[12] See, for example, "JTIDS/TIES Consolidate Tactical Communications," *EW*, September/October 1977.
MIL-STD-6016, "DoD Interface Standard Tactical Digital Interface Link (TADIL) J Message Standard," February 1997.
Stiglitz, M., "The Joint Tactical Information Distribution System," *Microwave Journal*, October 1987.
Toone, J., and S. Titmas, "Introduction to JTIDS," *Signal*, August 1987, pp. 55–59.

[13] U.S. Army Field Manual FM 24-41, "Tactics, Techniques, and Procedures for the Enhanced Position Location Reporting System (EPLRS)," Final Draft, July 1999.

[14] Mouly, M., and M. Pautet, *The GSM Systems for Mobile Communications*, Palaiseau: Cell and Sys, 1992.

4

Electronic Support

4.1 Introduction

Electronic support (ES) is the division of EW involving actions tasked by, or under the direct control of, an operational commander to search for, intercept, identify, and locate sources of intentional and unintentional radiated electromagnetic energy. The purposes of ES include immediate threat recognition and construction of an EOB. The main functions of ES are to produce operational intelligence, to provide steerage for EA, and to cue surveillance and target acquisition resources.

As illustrated in Figure 4.1, the major subdivisions of ES are [1]:

- *Search.* Before any other EW processes can be carried out, it is necessary to search for and classify electromagnetic signals of interest.
- *Intercept.* Signals of interest, once identified in the search process, are examined for their technical characteristics, such as bandwidth and modulation type, as well as their content, which may be monitored and recorded either by an operator or electronically.
- *Direction finding (DF).* The physical location of transmitters is identified by the DF process, based on steerage provided by the search process.
- *Analysis.* Analysis uses the information gained from the other ES processes to construct an EOB of the adversary, and attempt to infer the adversary commander's intent.

Traditionally, each of these processes has been carried out separately using its own special-purpose equipment. More recent technology makes

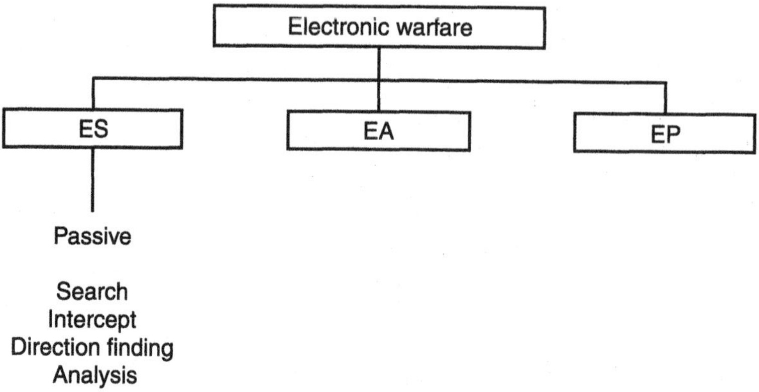

Figure 4.1 EW architecture.

possible the integration of two or more processes into a single receiver. The discussion below deals separately with each process so as to highlight its special characteristics and because, even if implemented in one equipment or detachment, there are still four distinct processes involved.

ES may target *adversary communications systems, adversary EA systems,* or *adversary electronics.* The electromagnetic emissions of communications systems of adversary communications systems are the primary traditional targets for ES, obtaining information for use in targeting and intelligence and supporting friendly EA. EA systems, like communications systems, emit electromagnetic radiation that can be exploited by ES. The targeting of adversary electronics other than communications systems is possible with specialized forms of ES equipment, although only over very short ranges. As a result of the short range, this type of target is accessible only on rare occasions.

ES functions, as collectors and processors of tactical information about an adversary, are closely related to other intelligence functions. This relationship is discussed in Chapter 6. In many cases, ES makes up the bulk (as high as 60–80%) of tactical information obtained about an adversary.

4.2 Search

Search involves reconnaissance of the electronic activity in the electromagnetic spectrum to classify the transmissions occurring within it. Search receivers must be placed within the coverage of the adversary systems to be detected and operated at a time during which transmissions are being made. In this sense, the search function can be seen as occurring in space, time, and frequency. The results of the search process provide the inputs for other EW functions. Searching can be conducted in general terms, or it can be made more specific by looking for particular call signs, types of modulation, or other signal or traffic characteristics. Sensitive, wideband receivers can be used to provide an indication of emitter characteristics.

The frequency range of interest can be searched by two main types of receivers: narrowband (or scanning) and wideband [2]. It is common for search receivers to be capable of operating across a wide range of frequencies. Operating frequencies stretching from below the HF band (around 1 MHz) to the middle of the UHF band (around 1 GHz) are not uncommon.

The theoretical minimum time required for a search receiver to detect an incoming signal is equal to the reciprocal of the receiver's bandwidth. Most practical receivers, however, require detection times much longer than this.

The search process is most commonly applied to the intended emissions of communications equipment. Under some circumstances, and at very short ranges, it is also possible to detect the operation of electronic equipment from unintentional radiation [3]. Protection against such detection is often referred to as TEMPEST [4], which is a form of EP.

The following sections describe the operation of major implementations of narrowband and wideband search receivers.

4.2.1 Narrowband Receivers

A *scanning receiver* is a single-channel receiver that can be rapidly stepped between a number of channels being monitored. At each step, the receiver determines whether the channel is active. Active channels may be flagged by the receiver pausing its scan to allow monitoring of the channel, or the receiver maintaining a list of currently active channels, which is made available to the operator in an appropriate manner. The narrowband search receiver can be seen as an automation of the traditional manual scanning carried out by operators to search for an adversary's transmissions.

The advantages of scanning receivers are that they are not limited to blocks of adjacent channels, and can step through an arbitrary collection of channels within their operating range; they are of low cost; and they can provide simultaneous decoding of signals.

The disadvantages of a scanning receiver are that it is effective in detecting only signals that are of duration longer than the time between visits to a channel, and are confined to a narrow portion of the frequency spectrum. Short-duration signals, particularly frequency hopping, and very wide-bandwidth signals, such as DSSS transmissions, may go undetected.

A scanning receiver is usually based on a superheterodyne receiver (Figure 4.2) that tunes to a single channel. Superheterodyne receivers are used because they provide a good tradeoff between sensitivity and selectivity. Scanning is achieved by regular stepping of the chosen frequency through the set to be searched.

The power received by the antenna is passed first into the RF stage. The functions of this first stage of the receiver are to select the channel of interest and to provide limited amplification of the received signal. The gain that can be provided in the RF stage is limited by the tradeoff between this gain and the bandwidth of the receiver. The higher the RF gain, the narrower the bandwidth. Unfortunately, providing any more than about 1,000 times amplification at this stage increases the selectivity of the RF stage to the point where it is attenuating not only out-of-band signals but also part of the wanted channel. The choice of channel is specified by the tuning control.

The output of the tunable band-pass filter is passed to the mixer, which converts the signal to a fixed intermediate frequency (IF) by mixing with a stable local oscillator frequency. In most modern receivers, the local oscillator includes a frequency synthesizer that enables accurate tuning to any frequency across a wide band. The signal is then further amplified and filtered before being passed to the detector. The purpose of the IF is twofold. First, it avoids the need for having a large gain at the RF stage, where there is a tradeoff between gain and selectivity. Second, the fixed frequency allows the optimization of the IF amplifier and filter to a single frequency, regardless of the frequency of the channel to which the receiver is tuned. The alerting system passes on information on active channels to an operator or directly to other ES processes, such as DF and intercept.

A scanning receiver may allow the operator to select one or more of the following parameters: frequencies to be searched; channel bandwidth;

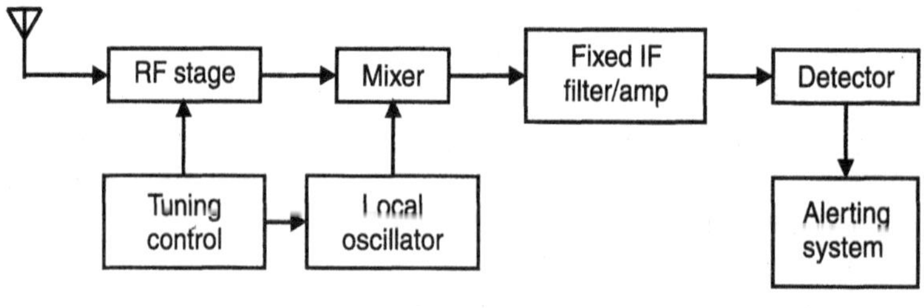

Figure 4.2 Simple block diagram of a superheterodyne search receiver.

the gain of the RF stage; and detection criteria, which may be power level or a particular type of modulation.

The detection time, which is the minimum time in which the scanning search receiver can detect a signal, is limited to the reciprocal of the channel bandwidth. The scan rate, therefore, can be no more than the channel bandwidth. In practice, scanning receivers do not achieve this limit. One reason for this is that a large number of false detections would occur as the scan rate approached this limit. Practical systems commonly achieve scan rates on the order of 20 channels per second.

4.2.2 Wideband Receivers

Wideband receivers are able to monitor a number of channels simultaneously. Therefore, they are more suitable for monitoring signals that are spread across several channels or are of short duration. Wideband receivers of the following types are used in searching for communications transmitters. These include a *channelized receiver, compressive receiver,* and *digital receiver.*

In addition, the following types of receiver are used in searching for other emitters, such as radar:

- *Crystal video receiver.* This type of receiver consists of a tuning stage to select the band of interest followed by a detector. Multiple receivers are often used in parallel to monitor a number of bands.
- *Instantaneous frequency measurement (IFM) receiver.* The IFM receiver allows the frequency measurement of pulsed signals, but can only respond to one signal at a time. The sensitivity of IFM receivers tends to be low because of the inherent high signal-to-noise ratio required to measure frequency.
- *Bragg cell receiver.* An instantaneous measure of the spectrum of received signals can be obtained using a Bragg cell receiver. This type of receiver employs a surface-acoustic-wave device to perform a Fourier transform. The output may be coupled via photodiodes and then to a charge-coupled device (CCD) that performs time averaging. The output from the CCD can then be displayed or read by a computer.

4.2.2.1 Channelized Search Receiver

As shown in Figure 4.3, a channelized receiver employs many fixed-tuning receivers in parallel to cover a complete band simultaneously. Although the frequency resolution of a channelized receiver may be lower than for a

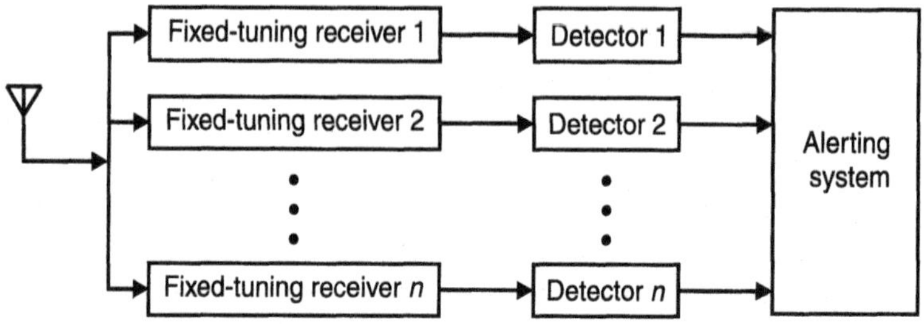

Figure 4.3 Channelized search receiver.

scanning receiver, it is generally good enough to ensure that a signal is detected. The receiver output may be used to direct a more detailed search.

A channelized receiver may search across a contiguous band, or may search a number of nonadjacent channels. Characteristics of the individual receivers and detection criteria may be chosen by the operator individually for each channel, chosen once by the operator for all channels, or fixed. Superheterodyne receivers may be used for each of the fixed-tuning receivers because they provide a good tradeoff between selectivity and sensitivity.

The channelized receiver searches all channels simultaneously. Any limitation on search rate comes from the alerting system. The properties of the individual receivers will, however, control the minimum time to detect a received signal. This minimum time is the reciprocal of the channel bandwidth of the individual fixed-tuning receiver. Practical systems will often have longer detection times to avoid a larger number of false detections.

4.2.2.2 Compressive Search Receiver

As shown in Figure 4.4, a compressive search receiver has a similar structure to the superheterodyne receiver. Wideband operation is obtained by replacing the tuning control with a sweep generator that covers an entire band. The rate of sweep can be very large, such that the bandwidth of the receiver lies within the bandwidth of the signal for much less than the detection time in any one sweep. The purpose of the compressive filter is to provide averaging over successive sweeps. This allows the compressive receiver to effectively monitor a whole band simultaneously, while maintaining high receiver sensitivity.

Because its receive frequency is controlled by a sweep generator, a compressive receiver can only be used to search across a contiguous band of frequencies. Practical limitations in the electronics used to implement the system limit the total bandwidth to approximately 1 GHz.

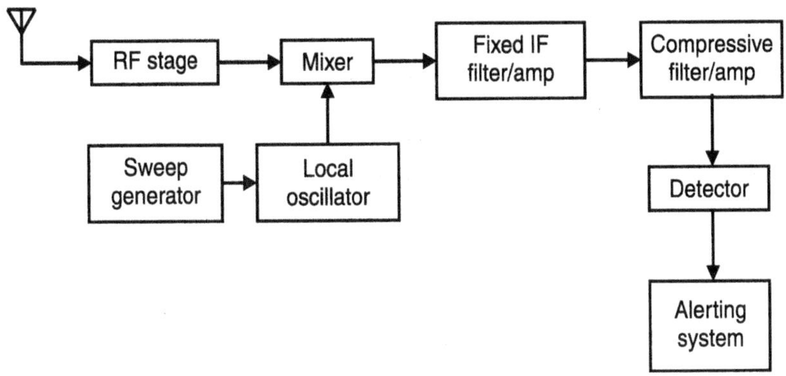

Figure 4.4 Compressive search receiver.

The search rate is limited by output display time, which can be no less than the reciprocal of the minimum resolvable bandwidth. This limitation is the same as has been discussed previously for other types of search receiver.

4.2.2.3 Digital Search Receiver

The digital search receiver, illustrated in Figure 4.5, consists of a receiver front end, an analog-to-digital (A/D) converter, and a signal processing system. The purpose of the receiver front end is to amplify the received signal to a level that can be handled by the analog-to-digital converter, to band-limit the incoming signal to prevent aliasing in the analog-to-digital converter, and possibly to shift the received signal to a known IF. Hence, the receiver front end would usually consist of an RF stage and a mixer. For HF and VHF receivers, the front-end receiver may be as simple as a low-pass filter.

The analog-to-digital converter converts the arriving signal to a digital form. The sampling rate of the converter must be at least twice the bandwidth

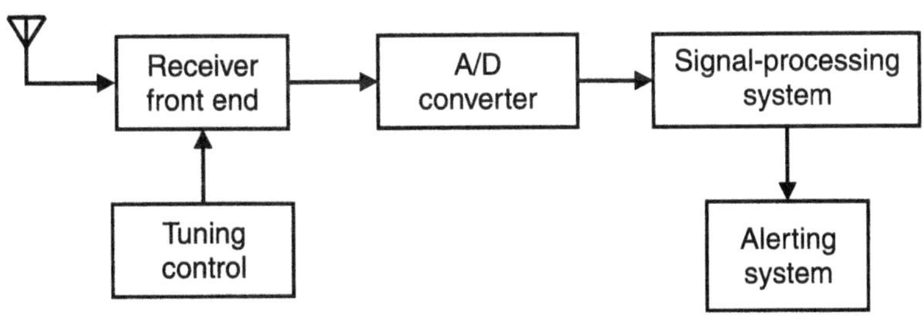

Figure 4.5 Digital receiver.

of the output signal from the receiver front end to prevent aliasing. The linearity of the analog-to-digital converter is critical to the performance of the whole system: any nonlinearity introduced during conversion will result in intermodulation distortion between received signals, potentially allowing a strong signal to obscure weaker signals.

The signal-processing system is responsible for the detection of incoming signals. A variety of algorithms may be used, but it is most likely that the processing will involve conversion to the frequency domain.

The required computational capacity required of the signal-processing system depends on the number of channels that are to be searched; whether or not it is necessary to continuously monitor these channels, or it is permissible to scan between them; the bandwidth of the individual channels to be searched; the sampling rate of the analog-to-digital converter; and the search algorithm that is used.

For a system where the processing is conducted using the fast Fourier transform (FFT), the channels are to be continuously monitored, and the sampling rate of the analog-to-digital converter is the minimum required to prevent aliasing, the required processing capacity expressed in operations per second can be calculated as follows.

If n channels are required, a $2n$-point FFT will be required. Each $2n$-point FFT requires $4n \log_2(2n)$ operations. With a sampling rate of f_s, the total number of operations per second (C) is then given by:

$$C = 2f_s \log_2 2n$$

If n channels are to be monitored, the channel bandwidth B is:

$$B = \frac{f_s}{2n}$$

The computational capacity required to monitor n channels is:

$$C = 4Bn \log_2 2n$$

Approximately 10^8 operations per second, with a sampling rate of 6 MHz, would be required to provide continuous search of 1,000 channels with a bandwidth of 3 kHz. This would enable detection of transmissions as short as 330 μs. This level of capacity is likely to be available in a system based on special-purpose, signal-processing hardware.

The number of channels searched could be increased by switching between banks of channels. While this would enable more channels to be searched without a significant increase in the computational capacity, the ability to detect short-term and rapidly changing signals is reduced, along with the ability to detect weak signals. Table 4.1 shows the computational capacity required for searching the whole HF and VHF bands.

4.2.3 Search and the Tactical Communications System

The search process aims to detect an adversary's transmissions. The application of search against the various subsystems of the tactical communications system is subject to a number of constraints, due to distance, terrain, and the use by the target system of directional antennas and other means of EP. The vulnerabilities of each tactical communications subsystem, along with the in-built protection against detection provided, are summarized in Table 4.2. Each subsystem is identifiable by its radio signature (i.e., type of transmission), including operating band, bandwidth, type of modulation, and traffic characteristics.

4.2.3.1 Tactical Trunk Subsystem

The employment of search to detect transmissions on terrestrial radio relay links between trunk nodes or command posts is hindered by the use of direction antennas and possibly other forms of active EP (including spread-spectrum communications) on these links. Figure 4.6 illustrates these difficulties. A search facility that is significantly off-axis, such as that labeled "1" in Figure 4.6, is unlikely to be able to receive transmissions from either

Table 4.1
Computational Capacity Required for Digital Search Receivers

Application	Minimum Sampling Rate	Number of Channels	Channel Bandwidth	C (Operations per Second)
Search VHF band 30–88 MHz	120 MHz	2,320	25 kHz	3×10^9
Search HF band 3–30 MHz	56 MHz	9,000	3 kHz	1.5×10^9
Search HF band in banks of 10 channels	56 MHz	900	30 kHz	1.2×10^9

Table 4.2
The Search Function and the Tactical Communications System

Tactical Communications Subsystem	Vulnerabilities	Protection
Trunk	Omnidirectional antennas in SCRA.	Directional antennas, long distance between transmitter and search facility, line-of-sight frequencies.
CNR	Omnidirectional antennas, short distance between transmitter and search facility, transmission only when messages sent.	Low power, low antennas, terrain screening.
Tactical data distribution	Omnidirectional antennas, short distance between transmitter and search facility.	Extensive EP.
Airborne	Height, omnidirectional antennas.	Only downlinks likely to be detected by tactical EW assets.

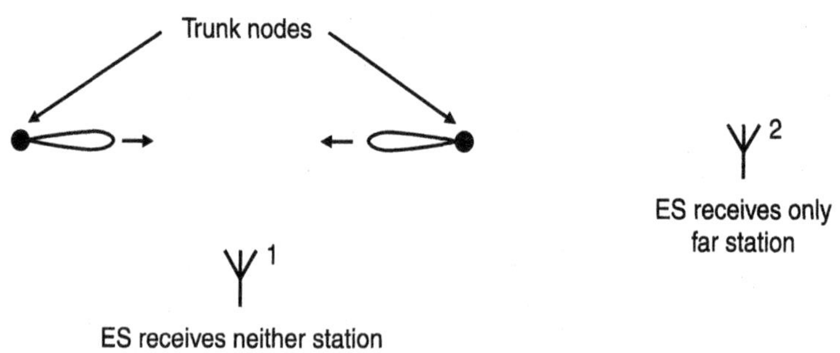

Figure 4.6 Difficulty of employing search against the trunk subsystem.

transmitter. A search facility that is aligned with the axis of the link (labeled "2") may be able to detect transmissions, but only from one end of the link.

The difficulty of using search against terrestrial radio relay links in the trunk subsystem is exacerbated by the greater distance that is likely between the search receivers and the trunk subsystem than is the case for CNR. This difficulty is greatest for ground-based search facilities, particularly when combined with the impact of terrain at VHF and higher frequencies. This is especially true for an on-axis search facility, such as the one labeled "2" in Figure 4.6. The impact of terrain may be reduced by the use of an airborne platform. However, it is unlikely to be possible to place an airborne search platform in a position from which it can intercept communications in both directions over a terrestrial radio relay link. The difficulty of detection of radio relay links is increased where additional EP techniques such as FH are employed.

The detection of an SHF down-the-hill link within a trunk node or command post poses similar difficulties to terrestrial radio relay. The difficulty is increased by the use of highly directional antennas and low transmission powers, especially for ground-based search facilities. Because of their natural protection due to position, SHF down-the-hill links are less likely to be protected by active EP than terrestrial radio relay links, which may reduce the difficulty of detection by an airborne search facility.

SCRA and CNRI are characterized by the use of line-of-sight frequencies, medium transmission powers, and omnidirectional antennas. The base stations for such systems are likely to be collocated with trunk nodes, meaning that they are also protected against detection by ground-based search facilities by distance and terrain. Mobile stations may be anywhere in the AO, often forward of the base station, and are therefore more vulnerable to detection than the base stations. This vulnerability is offset to some extent by the low height of antennas on mobile stations above the ground. Because the multiple channels supported by an SCRA base station tend to be separated from adjacent channels either by TDMA or FDMA, the detection of one transmission is likely to provide operating frequencies for all channels associated with that base station. In the past, SCRA base stations have not tended to employ EP for protection against detection, although future systems may reduce their signature by employment of measures such as FH.

Transmissions from trunk HF radio employ surface-wave communications where possible (e.g., on links between command posts within the AO) and sky wave otherwise (e.g., on rear links). Detection of surface-wave transmissions from trunk HF radio is possible anywhere within the coverage of the transmitter. For omnidirectional antennas, such as a vertical monopole or dipole, the area of coverage will be roughly a circle, the size of which depends on the permittivity and conductivity of ground and the transmitter power. Detection of sky-wave transmissions from trunk HF radio requires that the search facility be located in the footprint of the transmitter, as illustrated in Figure 4.7. It is unlikely that it would be possible to site a ground-based search facility within the AO capable of detecting transmissions

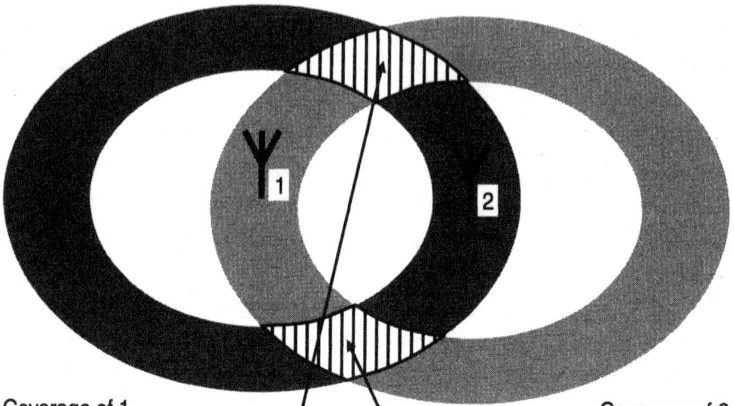

Figure 4.7 Footprint of sky-wave transmissions.

in both directions on a duplex link. This may be possible with an airborne platform. Search may be hindered by EP such as FH. However, HF FH tends to be slower than VHF (often only 10 hops per second), making it easier to track the changes in frequency.

The detection of an uplink associated with a high-capacity satellite trunk link is usually only possible if the search receiver is placed directly between the ground station and the satellite due to the high-gain antennas used. This may be possible using an airborne search facility, or even a ground-based search facility if the takeoff angle of the link is small enough. Downlinks for high-capacity trunk links are likely to be detectable over the whole AO. The use of EP, especially DSSS and FH, can considerably increase the difficulty of detection.

4.2.3.2 CNR Subsystem

VHF and UHF CNR employ mostly omnidirectional antennas and can be readily detected by search assets located within radio line-of-sight. HF CNR, like trunk HF communications, uses mostly surface-wave communications, and can be detected beyond radio line-of-sight, over an approximately circular area of coverage that depends on transmitter power and the type of ground. EP, mostly in the form of FH, may be employed to reduce the vulnerability to detection, especially by narrowband search receivers. The range over which VHF/UHF CNR can be detected depends critically on the height of the transmit antenna. Transmissions from higher-level headquarters, where antennas are likely to be mounted on masts, may be easier to detect than those from lower-level units even though the lower-level units may be located closer to the search facility. CNR transmitters are used further forward than trunk systems, reducing the likely distance between the transmitter and search facility.

4.2.3.3 Tactical Data Distribution Subsystem

Detection of transmissions from the tactical data distribution subsystem is hindered by its extensive use of EP, most likely including both DSSS and FH. If technology that can follow the FH is available, the use of omnidirectional antennas and the short distance between transmitter and search facility will aid detection. An airborne platform is likely to have some additional utility over a ground-based platform, although less than in the case of the tactical trunk subsystem, since terrain is unlikely to be the major limitation on detection of transmissions from the tactical data distribution subsystem.

4.2.3.4 Airborne Subsystem

An adversary's use of the airborne subsystem eases detection by elevating transmitters above the effects of terrain. As illustrated in Figure 4.8, an airborne ES system may be either below the airborne subsystem ("1"), in

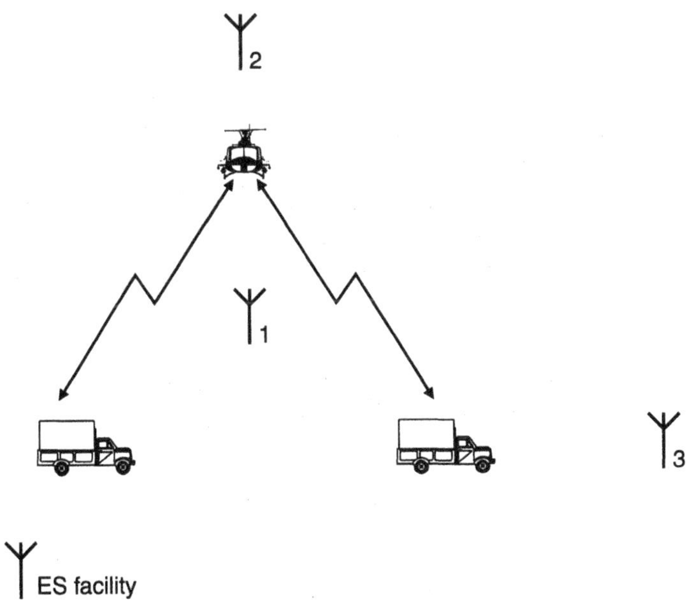

Figure 4.8 Employment of search against the airborne subsystem.

which case it will be able to detect both uplinks and downlinks; or above it ("2"), in which case it may only be able to detect uplinks. A ground-based search facility ("3") is likely to detect only the downlinks, unless it is placed close to one of the ground terminals. Some systems that conventionally use directional antennas, such as radio relay, may use omnidirectional antennas when mounted on the airborne platform, trading off their vulnerability to detection against the difficulty of maintaining communications from a moving platform.

4.3 Intercept (Monitoring)

Once the search function has identified a signal of interest, the signal is passed to an intercept receiver that further classifies an electromagnetic emission by its external characteristics, such as frequency, modulation, and bandwidth, and if possible, extracts the internal information content, often referred to as *monitoring*.

Some external characteristics, such as frequency and bandwidth of operation, may be relevant to the tasking of other EW assets. The intercept receiver settings, for example, could be transferred to DF receivers to obtain bearing samples of the transmitter. Others may be useful in deducing the adversary's EOB.

Fingerprinting refers to the process of identifying a transmitter by the unique characteristics of its spectrum, such as occupied bandwidth, tuning offset errors, and nonlinearities introduced in high-power amplifiers. In some cases, it may be possible to identify not only the type of transmitter but also an individual item of equipment. Ongoing fingerprinting can then track this item on the battlefield, thereby tracking its operating unit. In other cases, only the type of transmitter may be able to be determined. This may provide indications of the platform on which the transmitter is mounted. It may be known, for example, that a particular type of radio is used exclusively in armored vehicles. The transmitter bandwidth by itself may even provide useful information. Older VHF radios tend to transmit on a 50-kHz channel. Newer systems tend to use 25-kHz channels. It is likely that a unit using older equipment will have a lower level of readiness and training than a unit using the newer equipment.

In the past, recording of intercepted signals was always carried out manually by an operator. In modern systems, the recording process may be automatic or manual. Automatic systems are used for data and may be used for voice. For voice signals, manual gisting by an operator may take place in real time, even if automatic recording is available. This minimizes the delay in processing and dissemination of time-critical tactical intelligence. However, it cannot be carried out without operators proficient in the language used by the adversary.

Monitoring of an unencrypted net will reveal call signs, procedures, and possible locations as well as a wealth of tactical information. An operator proficient in the adversary's language will often gist a net in real time, passing information obtained to analysts for collation and fusing with collateral information. The use of encryption and other EP methods reduces the accessibility of internal information. Net activity can always be monitored in CNR even if the net is encrypted.

4.3.1 Characteristics of Intercept Receivers

Intercept receivers are similar to communications receivers [5], but because they are not designed to interoperate with one particular communications transmitter, they differ in a number of characteristics:

- *Sensitivity.* An intercept receiver should be able to receive and decode very weak signals. It would normally be expected that an intercept receiver would have a higher sensitivity than a communications receiver. This higher sensitivity is required because the intercept receiver will usually be farther from the transmitter than a communications receiver. Furthermore, if the transmitter uses a directional antenna, it will be oriented toward the intended receiver. This orientation will only rarely coincide with the direction to an intercept receiver. The sensitivity of an intercept receiver should be not less than the sensitivity of the supporting search receiver.
- *Noise performance.* Much better noise performance will generally be required in an intercept receiver than for a communications receiver. This flows partly from the requirement for higher sensitivity. The cost of improved noise performance may include larger size and weight, especially for very high-performance systems where special cooling may be required. The noise performance of an intercept receiver should also be not less than the sensitivity of the supporting search receiver.
- *Range of tuning.* An intercept receiver is likely to provide coverage of a much larger portion of the electromagnetic spectrum than a communications receiver. Tuning ranges of 10 kHz to greater than 1 GHz can be commonly found. A communications transceiver is

likely to cover a much narrower portion of the spectrum, such as 30 to 88 MHz for a VHF CNR.

- *Channel bandwidth.* An intercept receiver typically has a greater choice of channel bandwidth than a communications receiver. The channel bandwidth for a communications system is usually specified in the design of the equipment, and the operator has little or no control. VHF CNR, for example, usually has a fixed channel bandwidth of 25 kHz.
- *Granularity of tuning.* The tuning granularity of an intercept receiver will usually be much finer than a communications receiver. Tuning granularities as small as 1 Hz are often found, even in low-cost commercial equipment.
- *Modulation.* The modulation process carried out by a transmitter shifts the input baseband signal to another part of the electromagnetic spectrum. The information may be carried in the amplitude, frequency, or phase of the transmitted signal. A communications receiver usually has a fixed form of modulation, which is associated with the band in which it operates. For example, an analog voice radio would usually employ SSB in the HF band, while FM would be more common in the VHF band. An intercept receiver is likely to have a number of selectable modulation waveforms. For analog signals, these would include amplitude modulation (both single and double sideband) and frequency modulation. For digital signals, amplitude-shift keying (ASK), frequency-shift keying (FSK), and phase-shift keying (PSK) are likely to be available [6].
- *RF gain.* An intercept receiver may allow the operator to control a number of internal parameters, such as gain of the RF stage, that allow optimization of performance in marginal reception conditions.
- *Multichannel.* An intercept receiver may also allow more than one channel to be monitored simultaneously.

4.3.2 Intercept Receivers

The most commonly employed types of intercept receiver are the superheterodyne intercept receiver and the digital intercept receiver. This section describes the characteristics of these receivers, highlighting the likely differences from the employment of similar architectures as search receivers.

4.3.2.1 Superheterodyne Intercept Receiver

Figure 4.9 shows a simple block diagram of a superheterodyne intercept receiver. This block diagram is very similar to that shown previously for the superheterodyne search receiver, differing only in the use of the output. The operation of the individual blocks is somewhat modified by the unique characteristics of intercept receivers outlined previously.

The RF stage of an intercept receiver provides the same function as for a search receiver, providing amplification and selectivity. In an intercept receiver, it is likely that the operator will have explicit control over the gain-selectivity tradeoff. This may be particularly important when a very weak signal is intercepted and there is a strong signal on a nearby channel.

The local oscillator of an intercept receiver will almost certainly be based on a frequency synthesizer. Without this, the intercept receiver is unlikely to be able to tune across the wide range of frequencies required or provide sufficiently fine granularity of tuning. The use of a frequency synthesizer also greatly aids the accuracy, stability, and repeatability of tuning.

The mixer converts the incoming RF signal to a fixed IF. As for the search receiver, the purpose of the IF is twofold. First, it avoids the need for having a large gain at the RF stage, where there is a tradeoff between gain and selectivity. Second, the fixed frequency allows the optimization of the IF amplifier and filter to a single frequency, regardless of the frequency of the channel to which the receiver is tuned. The major additional feature of this stage in the intercept receiver is its flexibility: It must be able to deal with a variety of channel bandwidths and forms of modulation.

The detector provides the final stage of conversion of the signal to baseband. This stage also requires additional flexibility due to different forms of modulation. The output of the detector is passed to the recording system.

A superheterodyne intercept receiver would normally only allow a single channel to be monitored. If the monitoring of more than one channel is required, multiple receivers are used. The measurement of external characteristics may also be supported by a superheterodyne intercept receiver, possibly

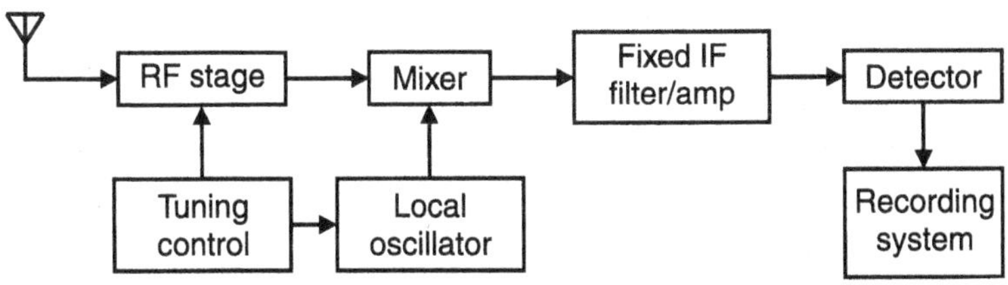

Figure 4.9 Simple block diagram of a superheterodyne intercept receiver.

by the connection of a spectrum analyzer to the output of the IF amplifier. A combined search/intercept superheterodyne receiver would add a frequency-scanning capability to the basic intercept receiver.

4.3.2.2 Digital Intercept Receiver

The block diagram of a digital intercept receiver is shown in Figure 4.10. The basic structure is the same as for a digital search receiver. In this case, however, the operation of the signal-processing system is quite different.

The receiver front end provides amplification, and filtering for selectivity and to prevent aliasing. The bandwidth of the output of the receiver front end may be lower for a digital intercept receiver than for a digital search receiver, thus reducing the computational load on the signal-processing system. This can be done without a compromise in performance, unless the intercept receiver is required to provide the continuous monitoring of multiple channels spread across a wide bandwidth.

The analog-to-digital converter converts the output signal from the receiver front end to a digital form. The sampling rate of the converter must be at least twice the bandwidth of this signal to prevent aliasing. As was the case for the digital search receiver, the linearity of the analog-to-digital converter is critical to the performance of the whole system.

The functions of the signal processing system in a digital intercept receiver are tuning (i.e., providing selectivity); measurement of external characteristics, such as channel bandwidth or modulation type; and demodulation, shifting signals to baseband and, if necessary, converting modulated digital signals into 0s and 1s.

The computational effort of performing the demodulation for one channel is likely to be higher than for the continuous search of the same channel in the digital search receiver. For this reason, there is some advantage in reducing the sampling rate of the analog-to-digital converter (i.e., provide some selectivity in the RF front end to reduce the bandwidth of the input

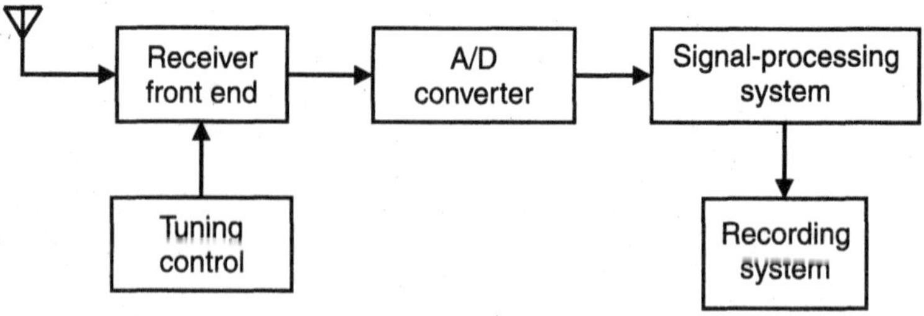

Figure 4.10 Simple block diagram of a digital intercept receiver.

signal to the analog-to-digital converter), or performing this down-sampling as the first step within the signal-processing system.

A digital intercept receiver may have an in-built recording and storage system. If a computer controls the receiver, this storage may be provided on the hard drive of the computer. Digital intercept receivers may also allow simultaneous monitoring of multiple channels, subject to the processing capacity of the signal-processing system and the rate at which the recording system can store the received data.

The difference between a combined search/intercept digital receiver and a conventional digital intercept receiver is mostly in the software used to control the radio.

4.3.3 Intercept and the Tactical Communications System

Intercept aims to monitor and record an adversary's communications. The growing use of embedded encryption in the tactical communications system is increasing the difficulty of decoding intercepted transmissions, especially at the tactical level. However, extensive information can still be obtained from traffic analysis of transmissions. Table 4.3 shows a summary of the interaction between the intercept process and the subsystems of the tactical communications system.

4.3.3.1 Tactical Trunk Subsystem

Before interception is possible, a transmission must be detected, which is difficult due to the use of directional antennas and the distance between transmitter and ES receiver. In addition to these difficulties, intercept is hindered by the universal employment of bulk encryption on terrestrial radio relay links. This means that the content of messages cannot be readily decoded and traffic analysis is severely constrained. Even where decoding is possible, it is almost certain that a ground-based intercept facility will intercept only one-half of a duplex link, which is of limited value. Even an airborne intercept platform may be unable to overcome this limitation for terrestrial transmissions, since it is unlikely to be able to locate itself so as to intercept both directions of a ground-based duplex link. The interception of other elements of the tactical trunk subsystem, including SHF down-the-hill links, trunk HF radio, and high-capacity satellite links is similarly constrained. Embedded encryption also hinders the interception of SCRA and single-channel satellite communications systems, although limited traffic analysis may be possible if transmission times are recorded, especially if multiple access is provided using FDMA rather than CDMA or TDMA.

Table 4.3
The Intercept Function and the Tactical Communications System

Tactical Communications Subsystem	**Vulnerabilities**	**Protection**
Trunk	Omnidirectional antennas in SCRA.	Directional antennas, long distance between transmitter and ES, line-of-sight frequencies, bulk encryption, continuous transmission, only one-half of full-duplex links accessible.
CNR	Omnidirectional antennas, short distance between transmitter and intercept facility, transmission only when messages sent.	Active EP: encryption and possibly FH.
Tactical data distribution	Omnidirectional antennas, short distance between transmitter and intercept facility.	Extensive EP, including encryption, continuous transmission.
Airborne	Height, omnidirectional antennas.	Only downlink likely to be intercepted.

4.3.3.2 CNR Subsystem

Interception and recording of CNR have traditionally been a major part of communications EW. Like search, intercept of these transmissions is aided by the use of omnidirectional antennas and the limitation of the employment of active EP to encryption and possibly FH. Because the CNR subsystem operates as a single-frequency, half-duplex, all-informed network, intercept tends to be more fruitful than in the trunk subsystem. For line-of-sight CNR operating in the VHF and UHF bands, the short distances between transmitters and ES facilities also favor interception. Some protection against intercept can be provided by employment of passive EP techniques such as terrain screening. The effectiveness of such techniques is greatest when only a single intercept station is used; it is rarely possible to choose transmitter sites that are screened from all possible locations in which an adversary could site a ground-based intercept facility. The employment of an airborne intercept facility greatly reduces the effectiveness of an adversary's terrain screening and other passive EP techniques.

Because of the almost exclusive use of unencrypted voice in CNR systems in the past, intercept has concentrated on monitoring the internal content of these transmissions. As well as providing a significant input to the intelligence process, intercepted transmissions also provided immediate tactical information of value to a commander. The replacement of voice with data transmissions for some applications does not change the utility of intercepted transmissions. However, the growing use of encryption for CNR at all levels is increasing the difficulty of decoding intercepted transmissions, whether voice or data. Even if encryption algorithms are weak, large-scale cryptanalysis is unlikely to occur at the tactical level, limiting intercept systems to acquiring external data to support traffic analysis.

4.3.3.3 Tactical Data Distribution Subsystem

Even if detection of transmissions from the tactical data distribution subsystem is possible, the extensive use of active EP, including DSSS, FH, and encryption, effectively prevents the decoding of transmissions at the tactical level. The use of TDMA, with each station usually transmitting in all of its allocated time slots, hinders the collection of external data that is useful for traffic analysis. An airborne intercept facility overcomes limitations of terrain, but offers no additional advantages for intercept in terms of overcoming the active EP of this subsystem.

4.3.3.4 Airborne Subsystem

Because of its elevation, the airborne subsystem is more vulnerable to intercept than the ground-based communications systems of which range it extends. The use of encryption, however, makes it unlikely that intercepted transmissions can be decoded. Because all transmissions passing through the airborne subsystem must travel on both an uplink and a downlink, the airborne subsystem may enable a ground-based intercept facility to intercept both halves of a duplex link even though it may not be able to monitor any of the ground stations.

4.4 Direction Finding

Direction finding (DF) can be used to provide information on the approximate location of emitters of electromagnetic radiation. DF is predicated on the basic principle of triangulation to find the position of an emitter. At least three DF receivers are positioned on a baseline, as illustrated in Figure 4.11 [7].

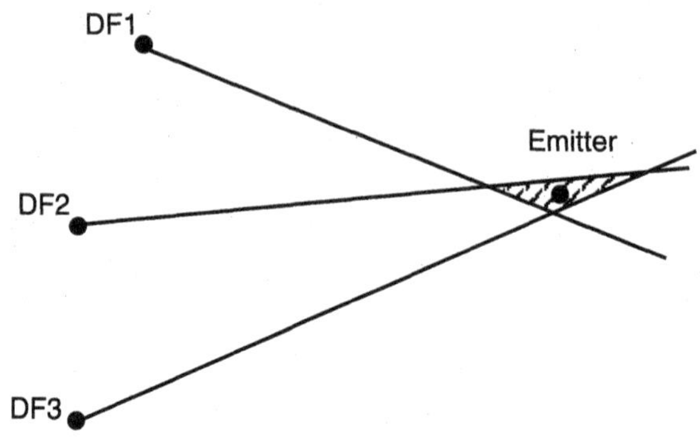

Figure 4.11 DF baseline and bearings taken on an emitter.

Each DF receiver has a special antenna, which is used to take a bearing toward the emitter. The bearings are plotted on a map, either manually or automatically, to form a triangle that should contain the emitter. The size of the triangle (and therefore the uncertainty in locating the transmitter) depends on the accuracy of each bearing.

At communications frequencies, the DF accuracies that are likely to be obtained are in the order of ±2° rms. At a range of 30 km, this leads to an uncertainty of ±1 km in the position of the emitter. Further insight into the location of the emitter can be gained by an analysis of the map to determine likely locations within the DF error. The accuracy of DF is sufficient for use in constructing an EOB. However, for long-range target acquisition DF is unlikely to have sufficient accuracy to be more than a cuing aid.

The error in estimating the range to a transmitter can be simply calculated for DF using a pair of DF receivers. Figure 4.12 shows how the relative error in estimating range varies with the angle between the lines of bearing and the error in these angles. For large angles between the lines of bearing, the relative error is equal to the error in the measurement, expressed in radians. For small angles between the lines of bearing, the error increases, passing 1 (i.e., 100%) when the angle between the lines of bearing reduces to less than twice the error in the measurement of bearings.

The use of a third DF station can reduce the error in the estimated location of the transmitter, and provide a check on the process by identifying when a very large error has occurred in the measurement of one of the bearings.

The bearing from a DF station to a transmitter may be determined by one of four methods: the *rotating directional antenna, amplitude difference,*

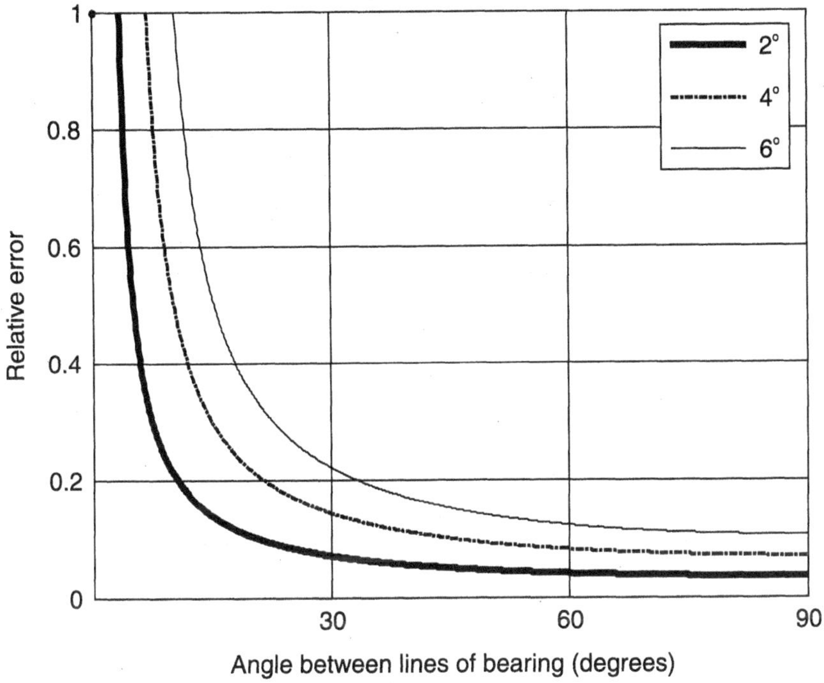

Figure 4.12 Relationship between range error, azimuth error, and angle between lines of bearing.

phase difference, and *time difference of arrival* (TDOA). The variation in output power with orientation of a rotating directional antenna can indicate the bearing to a transmitter. Alternatively, the bearing can be obtained by measuring the amplitude difference between the signals received at three or more antennas in an antenna array. The Watson-Watt and Wullenweber methods both use amplitude difference to obtain bearings. The phase difference between the signals received at antennas in an antenna array can be used to determine the bearing. Doppler and pseudo-Doppler DF systems use phase difference to calculate bearings. In systems that use pulsed transmission, the bearing to the transmitter can be obtained by measuring the time difference between the arrival of the pulse at two or more antennas. Such DF systems, while commonly used for radar DF, are not currently in common use in communications DF.

The advantages and disadvantages of the approaches commonly used in tactical communications DF are discussed in the following sections.

4.4.1 Sources of Error

DF is subject to a number of sources of error.

Equipment Error Typical, modern, high-performance DF equipment gives bearings with an accuracy of ±2°. Hand-held tactical units will often have less accuracy, possibly ±10°.

Short-Baseline Error If the baseline is such that the angle between bearing lines is less than approximately 45°, the triangle of error becomes significantly elongated. This leads to a large uncertainty in the location of the transmitter.

Cochannel Interference Most tactical DF systems are not capable of distinguishing between multiple received signals in a single channel. Where a significant level of cochannel interference is present, the DF will give erroneous bearings. The use of more than two DF locations may reveal errors caused by cochannel interference.

Adjacent-Channel Interference Strong signals in a channel adjacent to the one being DF'ed can lead to erroneous bearing.

Multipath Error Multipath error is a special case of cochannel interference. In this case, two or more received signals originate at the same transmitter but travel over different paths to the receiver. This can be caused by reflection from natural and man-made obstacles or by mixed modes of propagation, such as ground wave and sky wave. The choice of appropriate antennas can reduce the impact of mixed modes of propagation. Multipath error can often be minimized by the choice of an appropriate DF site, and is therefore often called site error.

Night Effect Night effect is a special case of multipath error that occurs when sky-wave propagation occurs at night but not during the day. This effect occurs at HF and lower frequencies, and is most relevant to DF of communications systems in the HF band.

Coastal Refraction Surface-wave propagation that crosses a coastline at an angle other than a right angle is subject to bending caused by refraction. This can lead to bearings that do not point at the transmitter. Coastal refraction is usually only significant at frequencies below 10 MHz.

Thunderstorms Thunderstorms can lead to erroneous bearings that point towards the storm rather than the monitored transmitter.

Rain Heavy rain can lead to significantly reduced received signal levels in the SHF and higher bands, reducing the range over which DF is effective.

4.4.2 Rotating Directional Antenna

A directional antenna transmits or receives power at higher levels in some directions than in others. The bearing to a transmitter can be sensed either by a maximum or minimum in received power as the orientation of the DF rotating directional antenna is changed. Two examples of radiation patterns of antennas that would be suitable for this task are shown in Figure 4.13. Figure 4.13(a) shows a radiation pattern of a directional antenna in which the bearing to a transmitter would be determined by a maximum in received power. Figure 4.13(b) shows a radiation pattern in which the bearing to a transmitter would be determined by a minimum in received power. The arrows indicate the direction in which a bearing to a transmitter would be obtained.

The accuracy with which the bearing can be determined by a rotating directional antenna depends on the width of the lobe or null on which bearings are based; the presence of other lobes or nulls that might give rise

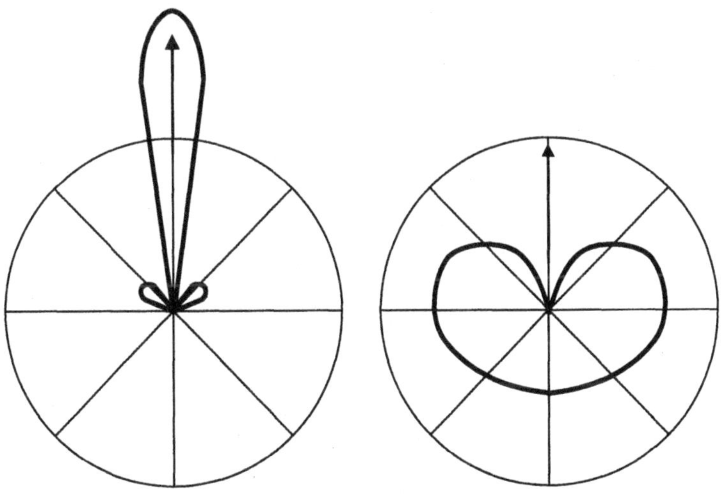

(a) (b)

Figure 4.13 Radiation patterns of directional antennas suitable for DF in which the bearing to a transmitter would be determined by a (a) maximum and a (b) minimum in received power.

to false bearings to phantom transmitters; and for weak signals, the sensitivity of the receiver.

Many antennas, such as the dipole, have radiation patterns that change markedly with frequency. This can affect both the relative gain and width of the lobes, and even the presence of the lobes and nulls in the radiation pattern. Ideally, the radiation pattern of the antenna should be constant across the frequency band of interest. In practice, some variation will always occur. For an antenna to be suitable for DF, it is necessary that the feature on which the bearing is based is present across the frequency band of interest and that the size of other features remains small enough that they do not interfere.

The advantages of a rotating directional antenna for DF are that it can distinguish between multiple transmitters on the same frequency, as long as the bearings to the transmitters are not too close together; and very small, hand-held, DF units can be constructed if the rotation is performed physically by the operator.

The use of a rotating directional antenna for DF has a number of disadvantages. Short-term transmissions, such as those from an FH transmitter, may not be reliably detected. The accuracy is limited by the width of the lobe or null of the antenna's radiation pattern on which bearings are based. Mechanical rotation is likely to be infeasible on many platforms, including aircraft and land vehicles on the move. Moving parts are likely to have high maintenance requirements. Antenna side lobes can lead to false peaks and nulls in power, although this can be minimized with careful antenna selection and construction.

4.4.3 Watson-Watt DF

The Watson-Watt DF method was developed to provide a DF system that is able to receive from all directions simultaneously, thus overcoming one of the principal disadvantages of the directional-antenna systems.

The basic principle of operation of the Watson-Watt DF system is illustrated in Figure 4.14, which shows a typical early implementation of the technique in which the bearing to a transmitter is displayed on a cathode ray tube (CRT). A crossed pair of loop antennas is used, with one antenna oriented north-south and the other east-west. Each antenna is connected to a receiver. The output signal level from the receiver is used to deflect the electron beam on the display, with the bearing indicated by the presence of a bright dot.

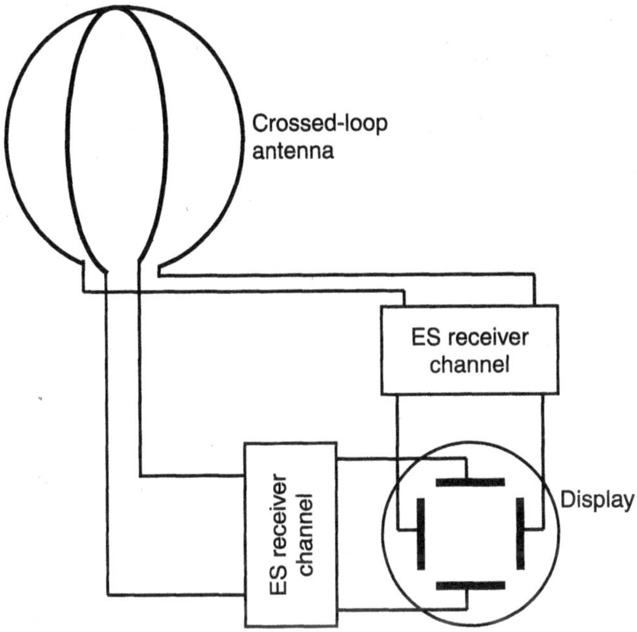

Figure 4.14 Watson-Watt DF system.

The arrangement shown in Figure 4.14 determines the bearing with an ambiguity of ±180°. A further antenna is required to resolve this ambiguity. A single dipole or monopole is typically used for this purpose.

A block diagram of a modern Watson-Watt DF system is shown in Figure 4.15. The CRT is replaced with a separate DF-bearing processor and DF-bearing display. The DF-bearing processor would often be implemented using digital signal processing. Both single channel (single receiver) and multichannel (multiple receivers) implementations are possible.

The sensitivity of a Watson-Watt DF system is affected by the characteristics of the antenna, the receiver, and the bearing processor.

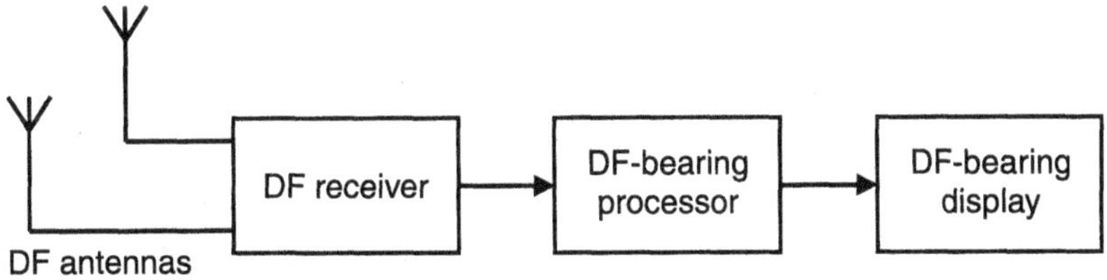

Figure 4.15 Watson-Watt DF system block diagram.

The advantages of a Watson-Watt DF system are its capability of receiving radiated power from all directions simultaneously, easy integration into mobile DF platforms, and the overall cost-effectiveness of the system compared to other methods. The disadvantages of a Watson-Watt DF system are the cochannel interference producing misleading bearings and the higher complexity and requirement for the calibration of the antenna system compared to other methods.

The description of the Watson-Watt technique was based on a loop antenna. Most such systems use either a loop or an Adcock antenna. The following sections examine the characteristics and relative advantages of each type of antenna.

4.4.3.1 Crossed-Loop Antenna

The simplest form of antenna for a Watson-Watt DF system is a crossed loop. This antenna is formed by placing a pair of loop antennas at right angles to one another, as shown in Figure 4.16. Each loop may be connected to a separate receiver, or a single receiver may be switched between the loops.

The received power from a loop antenna depends on the orientation of the loop with respect to the line to the transmitter. Maximum power is received from transmitters broadside to the loop; minimum power is received from transmitters end-on to the loop. Using a pair of crossed loops placed at right angles to one another, the bearing to a transmitter can be sensed by the relative power received from the two antennas. As stated above, the crossed-loop antenna provides bearings with an ambiguity of ±180°. This ambiguity can be resolved by the use of an additional antenna, sometimes called a *sense antenna.*

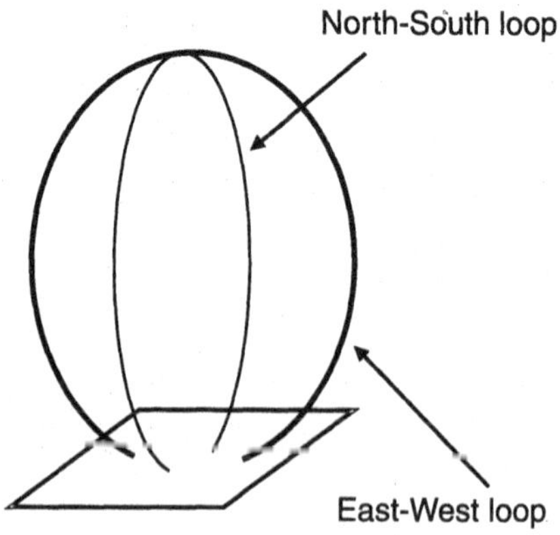

Figure 4.16 Crossed-loop antenna.

The primary disadvantages of the crossed-loop antenna are that it provides no capability to discriminate between horizontally and vertically polarized signals, and it is more susceptible than other types of antennas to cochannel interference due to mixed surface-wave and sky-wave propagation.

4.4.3.2 Adcock Antenna

The *Adcock antenna* is the most commonly used antenna for modern Watson-Watt DF systems. Figure 4.17(a) shows a plan view of two elements of an Adcock antenna; Figure 4.17(b) shows its radiation pattern. The two antennas may be implemented as monopoles or dipoles, as long as the two antennas are matched. Signals arriving from the left are inverted at the output of the summer; signals arriving from the right are passed through without being inverted.

The simplest practical Adcock antenna consists of four monopoles placed at the corners of a square. Larger numbers of elements of varying lengths may be used to increase the bandwidth. This system has the advantage that it can look in all directions simultaneously. With conventional processing, however, it can only operate in environments in which there is not significant cochannel interference.

An n-element Adcock is formed by placing a number of Adcock pairs around the dashed circle shown in Figure 4.17. The maximum diameter of this circle is half the wavelength of the highest frequency signal to be received. With four or more elements, a bearing to a transmitter with arbitrary azimuth can be determined. The bearing output of an n-element Adcock antenna, however, still contains an ambiguity of $\pm 180°$. This can be resolved by

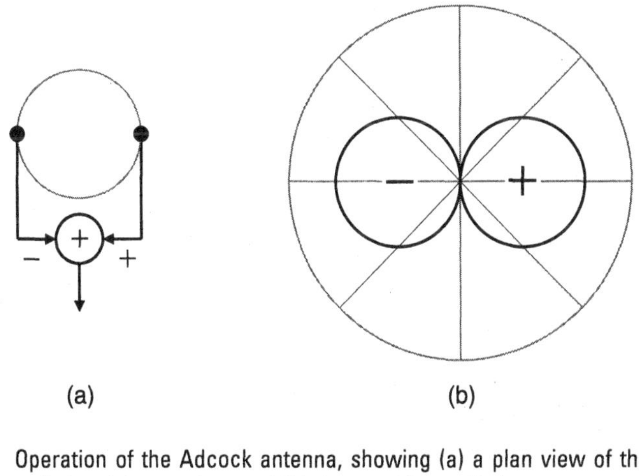

Figure 4.17 Operation of the Adcock antenna, showing (a) a plan view of the antennas and signal processing, and (b) the resulting radiation pattern.

comparing the phase of the output signals from the Adcock antenna with the phase of the output of an omnidirectional sense antenna. The sense signal will be in-phase for signals arriving from the right in Figure 4.17, and out-of-phase for signals arriving from the left. The sense signal can be obtained either by a separate antenna element placed at the center of the Adcock antenna or by summing the outputs of all *n* antennas.

The spacing between Adcock pairs is limited to half the wavelength of the highest frequency signal to be received. The length of the antennas is also limited by the highest frequency: half the wavelength for a dipole and a quarter of the wavelength for a monopole. These constraints limit the aperture of the Adcock antenna and its ability to suppress site error.

The major advantages of the Adcock antenna over the crossed-loop antenna are that it can discriminate between horizontally and vertically polarized signals, and it is less susceptible than other types of antenna to cochannel interference due to mixed surface-wave and sky-wave propagation.

4.4.4 Wullenweber DF

The Wullenweber DF system uses a circular array of antennas, as shown in Figure 4.18. A rotating commutator (known as a goniometer) selects the direction from which the antenna array receives. At any time, the output from a number of antennas is fed through variable delay lines to a sum/different receiver.

The Wullenweber system allows either a sum or difference output from the receiver to be used to determine a bearing. Like Watson-Watt DF systems, Wullenweber DF systems determine bearings based on the amplitude of received signals. For the sum display, the bearing is in the direction of the lobe. For the difference display, the bearing is in the direction of the null. Greater precision is typically obtained using the difference display.

Wullenweber DF systems tend to be large, static facilities. They are mostly used for HF sky-wave DF. The advantages of a Wullenweber DF system are its high resolution due to the use of more than two antennas in the array at any time, and the in-built flexibility in the choice of algorithms for determining bearings. The disadvantage of a Wullenweber DF system is that its large size makes it practical only for fixed installations, limiting its applicability to tactical DF.

4.4.5 Doppler DF

The Doppler DF system calculates bearings based on measured phase differences. The antenna for a Doppler DF system is usually a single element

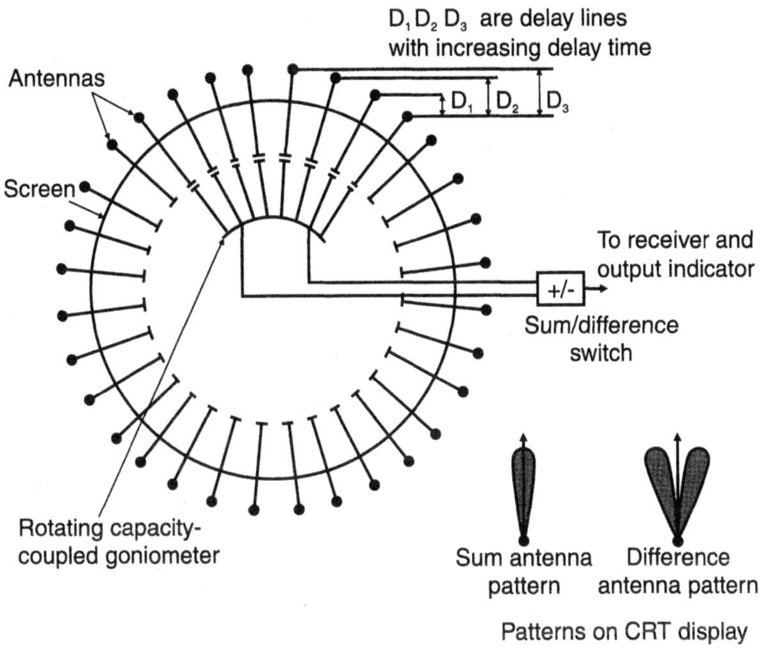

Figure 4.18 Wullenweber DF system.

mounted at the edge of a spinning circular plate. When the antenna is moving toward a transmitter, the received frequency increases; when the antenna is moving away from a transmitter, the received frequency decreases. The bearing to the transmitter can be determined by finding the bearings at which the maximum and minimum received frequency occurs. Doppler DF systems can distinguish between multiple transmitters sharing a single channel as long as the bearings from the DF antenna to the transmitters are sufficiently separated.

However, the requirement for physical rotation of the antenna has the same drawbacks as a DF system based on a rotating directional antenna. A pseudo-Doppler DF system replaces the rotating antenna with a circular array of fixed antennas and a fast-rotating commutator that switches the receiver's input between the antennas. This overcomes the requirement for a physical rotation and permits a much higher commutation rate than would be possible with physical rotation. The pseudo-Doppler DF receiver employs FM demodulation followed by phase comparison to the tone formed by the sequence of commutator angles, known as the *commutation tone.*

The advantages of a pseudo-Doppler DF system are that there is no inherent limitation on the aperture of the antenna, in contrast to Adcock antennas, which leads to a greater ability to suppress site error; multiple transmitters sharing a single frequency can be distinguished, as long as their bearings from the DF receiver are not too close; and it has antenna simplicity and wider bandwidth than Watson-Watt systems. The disadvantages of a pseudo-Doppler DF system are that the overall system complexity is typically higher than for Watson-Watt DF systems taking into account antennas, commutator and receiver; lower DF sensitivity, arising from the requirement for a high commutation rate to maximize DF sensitivity but a low commutation rate to minimize susceptibility to variation in the receiver group delay commutation rate; and the need to employ short antennas to minimize the impact of antenna reradiation.

4.4.6 The DF Baseline

The stations in a DF baseline may be deployed in a number of configurations: *standalone, convex, concave,* and *lazy-W.*

In standalone operation, a single DF station is used to provide a line-of-bearing to a target. Target location cannot occur using this mode, since only a single DF is employed. The standalone configuration is illustrated in Figure 4.19.

The convex configuration deploys the DF stations in a line, with those at the center closer to the adversary transmitter and those at the edge farther away. The convex configuration tends to be best when targets are located to the sides of the baseline; this is illustrated in Figure 4.20.

The concave configuration deploys DF stations with those at the edge of the baseline closest to the target. It is best used for targets that are directly in front of the baseline. The concave configuration is illustrated in Figure 4.21.

The lazy-W configuration is illustrated in Figure 4.22, and uses four or more DF stations to provide a good overall coverage where the approximate

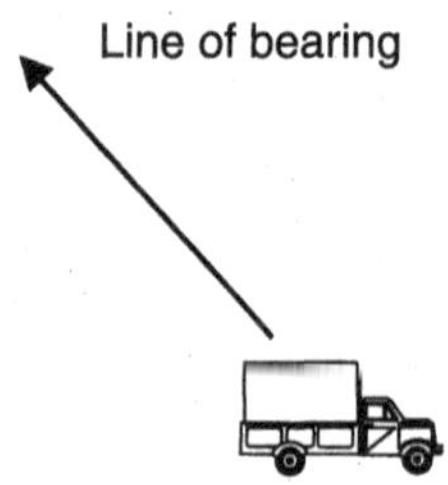

Figure 4.19 Standalone DF operation.

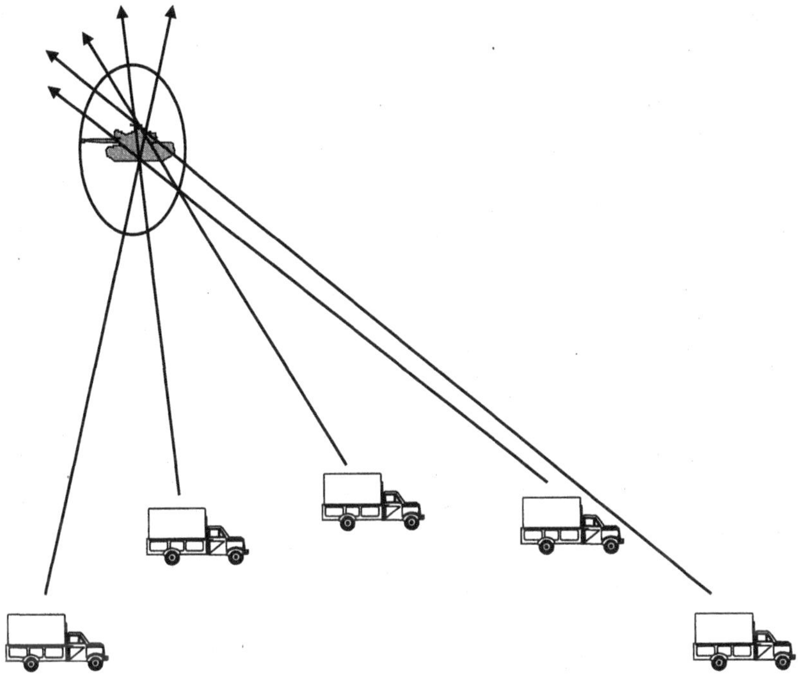

Figure 4.20 Convex DF baseline.

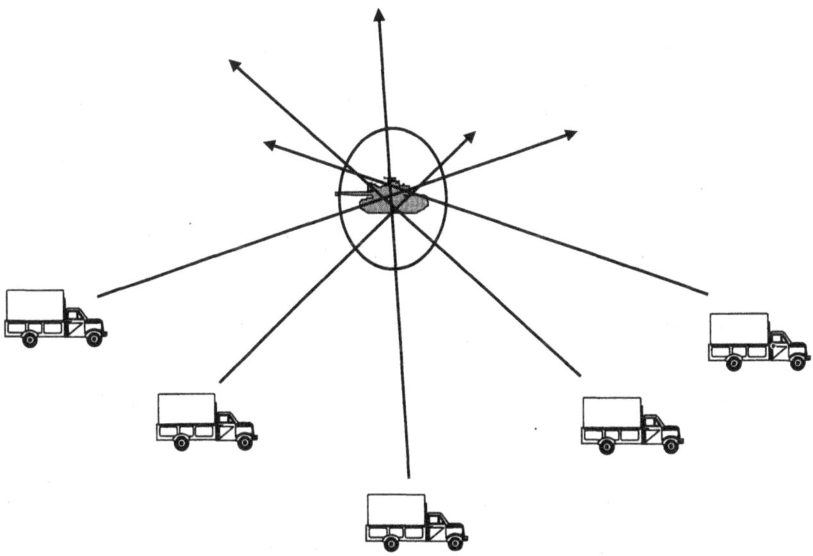

Figure 4.21 Concave DF baseline.

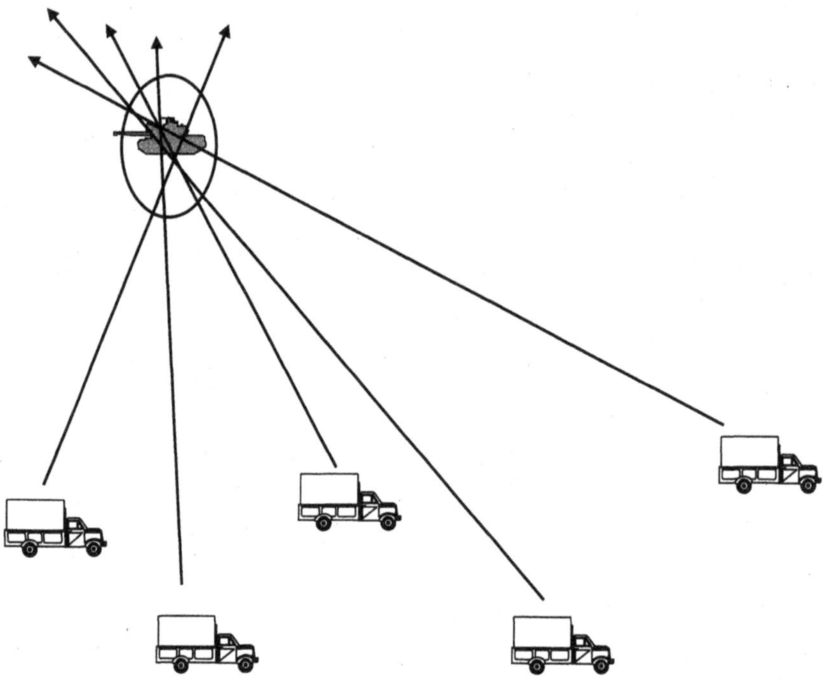

Figure 4.22 Lazy-W DF baseline.

locations of targets are not known in advance. This configuration may be easily extended to provide a larger area of coverage.

In practice, terrain and the tactical situation may place additional limitations on the locations of DF stations, making it difficult to achieve the desired configuration.

4.4.7 DF and the Tactical Communications System

The purpose of DF is to locate an adversary's transmitters, using the frequencies provided by the search process and prioritization from intercept. Issues associated with the employment of DF against the subsystems of the tactical communications system are very similar to those for search. The major difference is in the additional difficulties encountered by DF when targeting short-term and rapidly changing signals, including FH and burst transmission. Many ES systems have a search facility that hands off detected targets to a separate DF system for location. Detection, hand off, and DF (by all stations) must all occur before the target signal disappears. In practice, this means that search and DF systems must be highly integrated to operate against short-term and rapidly changing signals.

Table 4.4 shows a summary of the interaction between DF and the subsystems of the tactical communications system.

4.4.7.1 Tactical Trunk Subsystem

The use of DF against the tactical trunk subsystem suffers from the difficulties associated with search. Terrestrial radio relay links, SHF down-the-hill links, and satellite links are all protected by the use of directional antennas, transmission frequencies providing line-of-sight operation, and the long distances that will usually exist between transmitters and DF facilities. Even if a network of DF systems locates one or more nodes in the trunk network, this does not automatically lead to a picture of the network structure.

The use of DF on transmissions from SCRA base stations that are collocated with a trunk node may provide an indirect means of locating such a node, even though its main equipment (terrestrial radio relay and SHF down-the-hill links) may not be detectable. Used against SCRA mobile terminals, DF has the potential to reveal the locations of important personnel

Table 4.4
The DF Function and the Tactical Communications System

Tactical Communications Subsystem	Vulnerabilities	Protection
Trunk	Omnidirectional antennas in SCRA.	Directional antennas, long distance between transmitter and search facility, line-of-sight frequencies.
CNR	Omnidirectional antennas, short distance between transmitter and search facility, transmission only when messages sent.	Low power, low antennas, terrain screening, possibly FH.
Tactical data distribution	Omnidirectional antennas, short distance between transmitter and search facility.	Extensive EP, including FH.
Airborne	Height, omnidirectional antennas.	Only downlinks are likely to be detected by tactical EW assets.

(including commanders) operating outside command posts. Ground-based DF of base stations will often be easier than for mobile stations due to the placement of base-station antennas on masts and mobile-station antennas close to the ground. The height of the base-station antennas will reduce the shielding effect of terrain.

DF of long-range trunk HF systems within the AO is likely to yield only the location of the far end of the link, unless the DF stations are sufficiently close to the transmitter to receive surface-wave signals. For many uses of these systems, such as those providing a rear link, the location of the station in the AO is of much more interest than the station outside the AO.

4.4.7.2 CNR Subsystem

The application of DF against the CNR subsystem has perhaps the greatest potential to reveal an adversary's EOB, regardless of whether VHF or HF CNR is used. The use of omnidirectional antennas and the short distances between transmitter and DF facility hinders the use of terrain screening, while the use of net structures tied to command arrangements eases the analysis task.

Short-term and rapidly changing signals (such as burst transmission and FH, which are becoming more commonly used as a means of EP in the CNR subsystem) provided a particular challenge to DF. Effective DF requires that the target is detected and passed to the DF system and the line-of-bearing is acquired all within the time during which transmission occurs in a single channel. To be effective against such signals, it is necessary to have a high level of automation and integration between search and DF, since the involvement of a human operator in the process will lead to response times that are far too long. One possible approach is to have a number of integrated search/DF facilities that store a line-of-bearing, reception time, and frequency for each detected signal. This data can then be passed to an analysis facility that combines lines-of-bearing associated with the same frequency and reception time to locate the transmitter.

4.4.7.3 Tactical Data Distribution Subsystem

While DF of the transmitters in the tactical data distribution subsystem is assisted by the use of omnidirectional antennas, the short distance between transmitter and DF facility, and the regular transmissions of many stations to share situational awareness data, the extensive use of EP including FH, DSSS, and burst transmission provides a high level of protection for transmitters. Therefore, with currently available technology DF against the tactical data distribution subsystem is very difficult. Because every vehicle in a mechanized force is likely to be a node of this subsystem and to transmit regularly (perhaps more than once per minute), DF has the potential to reveal completely the force's disposition if the impact of the subsystem's EP can be overcome.

4.4.7.4 Airborne Subsystem

DF applied to the airborne subsystem is most likely to reveal only the location of the airborne platform, unless the DF system can also be targeted against the associated ground terminals. DF may, however, be useful in identifying the source of transmissions as an airborne platform because of the high concentration of emissions that are likely.

4.5 Analysis

Analysis is carried out in an attempt to put together a comprehensive adversary EOB and to obtain other information that may give insight into an adversary commander's plans or intent. Information for analysis might be obtained from the signal characteristics; traffic patterns in transmissions; information content, if decoding and decrypting are possible; and transmitter location [8].

Information from these sources is analyzed to provide an overall picture of the adversary's deployment. Conclusions are usually drawn about activity, future intentions, headquarter locations, unit types, and formation boundaries. However, as with any intelligence, care must be taken to ensure that the conclusions are tested against collateral (other intelligence) sources, to counter possible enemy deception techniques.

4.5.1 Traffic Analysis

In EW, traffic analysis deals with the study of the external characteristics of radio communications for the purpose of obtaining information on the organization and operation of the communications system. This information is then used to construct an EOB, which shows the disposition and command structure of the adversary's forces. In some circumstances it may also be possible to identify the adversary commander's intentions, even without being able to read the contents of transmissions [9].

Traffic analysis uses three basic types of data. *Intercept data* is the contents of the messages transmitted. This is sometimes subdivided to distinguish information gained from procedural parts of the messages, such as call signs, from the actual contents of the messages passed. *External data* includes the time and frequency of a transmission. *Collateral information* may be available for a particular transmission, such as DF bearings to locate the transmitter or information available from other sources of intelligence.

4.5.2 Cryptanalysis

Cryptanalysis is the process of decoding an adversary's encrypted data. There are usually insufficient time, resources, and expertise at the tactical level to perform cryptanalysis, except possibly for very simple, hand-driven codes. Cryptanalysis is usually carried out at the strategic level, where it is a key function of SIGINT [10]. Encrypted data that is intercepted by tactical ES units may be passed to a higher level for cryptanalysis.

4.5.3 The Analysis Function and the Tactical Communications System

Analysis provides the means of fusing data obtained from a variety of ES sources to provide a higher level of information about an adversary. The characteristics of the subsystems of the tactical communications system are closely related to the types of information that can be obtained. Table 4.5 shows a summary of the interaction between DF and the subsystems of the tactical communications system.

Table 4.5
The Analysis Function and the Tactical Communications System

Tactical Communications Subsystem	Vulnerabilities	Protection
Trunk	Network topology may be inferred.	Bulk encryption prevents traffic analysis.
CNR	Traffic analysis.	(Possibly) FH.
Tactical data distribution	Large number of transmitting stations.	TDMA prevents traffic analysis.
Airborne	Transmissions opened up to traffic analysis.	Only downlinks likely to be available for analysis by tactical EW assets.

4.5.3.1 Tactical Trunk Subsystem

Analysis of ES data obtained from the tactical trunk subsystem may be used to build up a picture of the structure of a terrestrial radio relay network. This network structure may reveal weak points at which the network may be easily attacked. It may also be possible to infer the approximate locations of command posts from the network structure. The use of bulk encryption and continuous transmission in the trunk subsystem, however, reduces the external characteristics of the transmissions to the point that they convey little or no information about the structure of the traffic being carried.

4.5.3.2 CNR Subsystem

Analysis associated with the CNR subsystem is primarily concerned with traffic analysis. The CNR subsystem is characterized by a number of features that make possible traffic analysis by an adversary: transmissions occur only when a message is to be sent; the hierarchical structure of nets in the CNR subsystem matches the command structure of the force; the level of activity on the net will tend to reflect the intensity of operational activities; and not all transmissions are encrypted, especially at battalion and below.

For encrypted nets, the information available for traffic analysis includes the start time and duration of transmissions, the frequency on which transmissions are made, and, if DF is available, the physical location from which a transmission is sent.

Stations that transmit on the same frequency are almost certainly associated with the same net. For a command net, it is likely that the net control station is collocated with the superior headquarters and that it will be the station that either transmits or receives the most messages. The net control station is also the most likely station to initiate a check of communications, which will see all the other stations reply in sequence. It is also the station that will take the lead in fixing problems, such as when cryptographic equipment loses synchronization. Other types of nets, such as artillery nets and EW nets, will also tend to have distinguishing characteristics as a result of their operational procedures. Collocation of stations on multiple nets also gives information on the command relationship between organizations.

For nets that are not encrypted, traffic analysis may also make use of some internal characteristics of a transmission, such as station call signs. Even where suitable encryption equipment is available, judicious application of jamming may force an adversary to drop back to unencrypted voice operation. Traffic analysis of unencrypted nets is aided by an adversary who does not take special measures to conceal the connection between call signs, nets, and the structure of the force. Examples include using call signs that were based on abbreviations of the names of units (as was done by the British for a period during World War I), using the same call sign for a station on more than one net, using the same call sign for a station on both primary and alternate frequencies on the same net, basing the selection of call signs on a theme (such as the names of flowers) that can be identified with a particular unit or formation, and utilizing the same frequencies and call signs for a net for long periods of time.

4.5.3.3 Tactical Data Distribution Subsystem

The use of a variety of EP techniques and a regular structure of transmission in the tactical data distribution subsystem also removes structure that might be used for traffic analysis. Little information is likely to be available about command arrangements, because every station in a formation appears to be talking to every other station.

4.5.3.4 Airborne Subsystem

The concentration of transmitters in the airborne subsystem and its use for range extension remove any terrain screening that might have been obtained by the siting of ground-based transmitters. For CNR, this has the potential to open up to traffic analysis nets that might not otherwise have been intercepted.

4.6 ES Platforms

ES can be conducted from ground-based or airborne platforms. Airborne ES may complement or provide an alternative to ground-based systems. Aircraft have several advantages, including increased intercept range; a shorter reaction time, allowing critical information to be passed to tactical users in real time; greater mobility and flexibility, permitting deployment over any terrain in response to urgent requirements; and enhanced survivability by operating in a standoff role.

4.6.1 Ground Platforms

Ground-based ES may operate from ground vehicles or in a manpack role.

4.6.1.1 Ground Vehicles

Ground vehicles, including small four-wheel drive vehicles, trucks, and armored vehicles, are commonly used as ES platforms. The type of vehicle is usually dictated by the type of force being supported; ES assets require at least the same mobility as other parts of the force. Therefore, if the force is mounted in armored vehicles, it is usually necessary to mount the supporting ES in armored vehicles.

The weight that can be carried by a ground ES platform varies with the type of vehicle, ranging from approximately 500 kg for a small four-wheel drive vehicle to in excess of 1,000 kg for a truck. Many ground ES platforms are mounted in shelters, which may weigh significantly more than the equipment that they house. Space may also be a significant limitation in some platforms, especially armored vehicles.

Vehicle-based operation permits ES antennas to be deployed on masts, allowing operation up to approximately 30m above the ground. Effective ranges at VHF and higher frequencies are likely to be approximately 15 to 50 km, depending on terrain and siting.

Operation of ground vehicle-based ES on-the-move may not always be possible. It is unlikely that masts longer than 5m could be deployed on a moving ground vehicle; in fact, mast height would probably need to be much less for some types of antennas or for movement through foliage. Significant additional errors are likely to be caused to DF, both because of higher errors in measurements of bearings and less precise knowledge of the location from which the bearings were taken. The moving vehicle is likely to move in and out of coverage of targets, limiting its effectiveness, especially in search and intercept operations. It is unlikely to be possible to deploy HF sky-wave antennas on a moving vehicle due to their size. A human operator inside a moving ground vehicle is unlikely to be able to work effectively, especially if writing or using a keyboard is required.

4.6.1.2 Man-Portable ES

The major differences in characteristics between ES systems designed for vehicle-based operation and those designed to be man-portable are the lower allowable weight of the man-portable system, the likely accuracy of bearings obtained in a man-portable DF system, and the height at which both ES and communications antennas can be deployed.

With the development of digital signal processing (DSP) technology, the impact of weight on search, intercept, and DF performance is reducing. However, the establishment of a man-portable ES command system is likely to be more problematic. This is due to the weight and bulk of the computing and communications systems that are required.

4.6.2 Air Platforms

Air platforms overcome the limitations of ground-based ES, using their height to provide long-range line-of-sight paths. These air platforms may be manned or unmanned fixed-wing aircraft or helicopters, UAVs, or aerostats. The payload weight supported by these platforms ranges from approximately 10 kg in a tactical UAV to in excess of 10,000 kg in a transport aircraft.

As discussed in Chapter 2 for airborne communication systems, in the VHF and higher bands, deployment of an airborne ES platform at an altitude of 65,000 ft would provide coverage of an area 500 km in diameter. At this height, highly directional antennas may be required to reduce the impact of cochannel interference. The area of coverage can be increased by the use of two or more airborne platforms, possibly with direct communications links to facilitate coordination and exchange of data. An airborne platform also has advantages for ES in the HF band, especially for sky-wave transmissions. A ground-based ES platform would usually have to be deployed within the coverage of both ends of a communications link. This may be difficult to achieve for a ground platform, but is easily achieved for an airborne platform. An airborne DF platform can provide its own baseline by measuring bearings to a transmitter from different locations as it moves over the battlefield. This approach will not be effective against short-term or rapidly changing transmissions, since it relies on the continued presence of the signal.

An airborne ES platform may also carry EA assets.

Endnotes

[1] U.S. doctrine on ES is covered in:
U.S. Army Field Manual FM 34-1, "Intelligence and Electronic Warfare Operations," September 1994.
U.S. Army Field Manual FM 34-2, "Collection Management and Synchronization Planning," March 1994.
U.S. Army Field Manual FM 34-36, "Special Operations Forces Intelligence and Electronic Warfare," September 1991.
U.S. Army Field Manual FM 34-37, "Echelons Above Corps (EAC) Intelligence and Electronic Warfare (IEW) Operations," January 1991.
U.S. Army Field Manual FM 34-40-9 "Direction Finding Operations," August 1991.

[2] The characteristics of search receivers can be found in:
Neri, F., *Introduction to Electronic Defense Systems,* Norwood, MA: Artech House, 1991.
Schleher, D. C., *Electronic Warfare in the Information Age,* Norwood, MA: Artech House, 1999.
Vaccaro, D. D., *Electronic Warfare Receiver Systems,* Norwood, MA: Artech House, 1993.

[3] See, for example, Wright, D., *Spycatcher,* Richmond, VA: William Heinemann, 1987.

[4] *Electromagnetic Pulse (EMP) and TEMPEST Protection for Facilities,* Engineering and Design Pamphlet EP 1110-3-2, U.S. Army Corps of Engineers, December 1990. A number of other publications have recently been declassified, including:
NACSIM 5000, "TEMPEST Fundamentals," National Security Agency, February 1982.
NACSEM 5112 (RP-4), "NONSTOP Evaluation Techniques," National Security Agency, April 1975.
NSTISSI No. 7000, "Tempest Countermeasures for Facilities," National Security Agency, September 1993.
NSTISSAM TEMPEST/2-95, "Red/Black Installation Guidance," National Security Agency, December 1995.

[5] The characteristics or intercept receivers are discussed in, for example, Rohde, U. L., and T. T. N. Bucher, *Communications Receivers: Principles and Design,* New York: McGraw Hill, 1976.

[6] See, for example, Sklar, B., *Digital Communications,* Englewood Cliffs, NJ: Prentice-Hall, 1988; or Proakis, J. G., *Digital Communications,* 2nd Edition, New York: McGraw-Hill, 1989.

[7] A large number of books have been written on DF, including:
Cotter, C. H., *The Principles and Practice of Radio Direction Finding,* London: Pitman, 1961.
Gething, P. J. D., *Radio Direction Finding and Superresolution,* 2nd Edition, Stevenage, Herts: Institution of Electrical Engineers, 1991.
Lo, Y. T., and S. W. Lee, *Antenna Handbook: Theory, Applications, and Design,* New York: Van Nostrand Reinhold, 1988.
Watson, D. W., *Radio Direction Finding,* New York: Van Nostrand Reinhold, 1971.

[8] Discussions of the techniques of analysis can be found in:
Callimahos, L. D., and W. F. Friedman, *Military Cryptanalytics,* Laguna Hills, CA: Agean Park Press, 1985.
Levite, A., *Intelligence and Strategic Surprises,* New York: Columbia University Press, 1987.
Sinkov, A., *Elementary Cryptanalysis: A Mathematical Approach,* New York: The Mathematical Association of America, 1966.

[9] Callimahos, L. D., and W. F. Friedman, *Military Cryptanalytics,* Part 2, Vol. 2, Laguna Hills, CA: Agean Park Press, 1985, pp. 477–486.

[10] Many books have been written on the contributions of cryptanalysis, especially in World War II, including:
Smith, M., *The Emperor's Codes: Bletchley Park and the Breaking of Japan's Secret Ciphers,* London: Bantam Press, 2000.
Smith, M., *Station X: The Codebreakers of Bletchley Park,* London: Channel 4 Books, 1998.

5

Electronic Attack

5.1 Introduction

Electronic attack (EA) is the division of EW involving the use of *electromagnetic energy* to attack personnel, facilities, or equipment with the intent of degrading, neutralizing, or destroying adversary combat capability. In a similar manner to indirect fire, EA aims to minimize the effect of adversary devices that rely upon the EM spectrum [1].

The subdivisions of EA are illustrated in Figure 5.1. Jamming aims to impair the effectiveness of the adversary's electronic equipment or systems by degrading the quality of the signal at a receiver. Electronic deception aims to confuse or mislead the adversary or the adversary's electronic systems. Neutralization is the use of electromagnetic energy to either disrupt or permanently damage adversary communications or electronic equipment. The power required for neutralization is typically many times larger than that required for the effective jamming of a receiver. Neutralization will be covered briefly in this chapter. RF directed energy weapons, a major new means for achieving neutralization of communications and other electronic systems, are examined in more detail in Chapter 7.

EA can target adversary communications systems through jamming, deception, and neutralization; adversary ES through jamming, deception, and neutralization; and adversary electronics, primarily through neutralization.

EA does not exist in isolation, but rather as part of a force's fire plan, which in turn is part of the operational plan. The relationship between EA and other parts of the operational plan is discussed in Chapter 6.

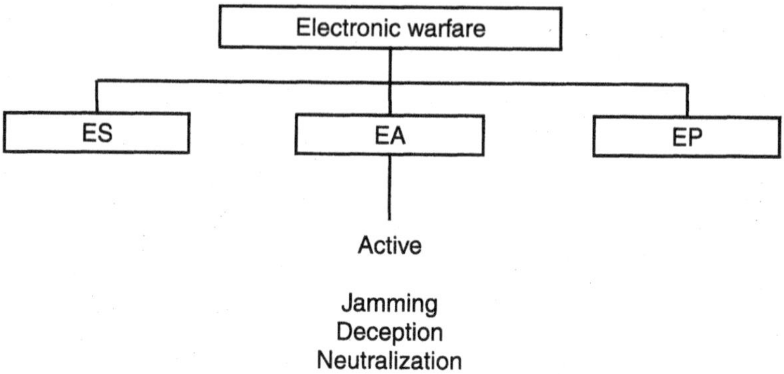

Figure 5.1 EW architecture.

ES provides a variety of information required for EA, including frequencies to be used. While EA is being carried out, ES also provides monitoring of the effectiveness of the EA.

5.2 Jamming

The aim of communications jamming is to disrupt adversary communications by delivering more power at the receivers in the net than is delivered by the intended transmitter. Receivers, not transmitters, are jammed. Effective jamming requires the jammer either to be closer than the intended transmitter, which is unlikely in most tactical situations, or to transmit on higher power settings. Either of these approaches makes the jammer vulnerable to adversary ES and weapon systems. Therefore, the jammer is normally very mobile and does not stay in the same position for very long after jamming. Ground-based jammers also tend to operate in pairs, one jamming while the other is moving. Continuous jamming can therefore be achieved while minimizing opportunities for detection.

Jamming may aim to jam all the receivers associated with a particular part of the adversary's communications system (e.g., a net), or it may aim to jam a single receiver. Jamming is normally used with ES support, so that the effectiveness of jamming can be determined. The supporting ES may be remotely located or collocated with the jammer.

Jamming can be used to target either adversary communications systems or adversary ES. Jamming used to reduce the effectiveness of adversary ES is often referred to as masking. The use of masking allows friendly communications systems to operate with reduced risk from adversary ES. Jamming is not used against electronics other than RF receivers.

5.2.1 Jamming-to-Signal Ratio

The performance of a jammer depends on the relative signal levels from the jammer and the communications transmitter at the input of a receiver being jammed. It is therefore a receiver that is jammed, not a transmitter.

The power received by the receiver's antenna from the transmitter depends on the power generated by the transmitter's power amplifier P_T, the loss in the feeder system to the transmit antenna L_T, the gain of the transmit antenna in the direction of the receiver G_{TR}, the propagation loss due to the path between the transmitter and receiver L_{TR}, and the gain of the receiver's antenna in the direction of the transmitter G_{RT}. With the losses and gains expressed as power ratios, the power received from the transmitter P_{RT} can be expressed as:

$$P_{RT} = \frac{P_T \times G_{TR} \times G_{RT}}{L_T \times L_{TR}}$$

The power received by the receiver's antenna from the jammer depends on the power generated by the jammer's power amplifier, which lies within the bandwidth of the receiver P_J; the loss in the feeder system to the jammer's transmit antenna L_J; the gain of the jammer's transmit antenna in the direction of the receiver G_{JR}; the propagation loss due to the path between the jammer and receiver L_{JR}; and the gain of the receiver's antenna in the direction of the jammer G_{RJ}. With the losses and gains expressed as power ratios, the power received from the jammer P_{RJ} can be expressed as:

$$P_{RJ} = \frac{P_J \times G_{JR} \times G_{RJ}}{L_J \times L_{JR}}$$

These factors are illustrated in Figure 5.2.

The jamming-to-signal ratio (JSR) can be written as:

$$JSR = \frac{P_{RJ}}{P_{RT}}$$

Two further factors are involved in the transfer of power into the receiver, but both of these impact equally on signals from transmitter and

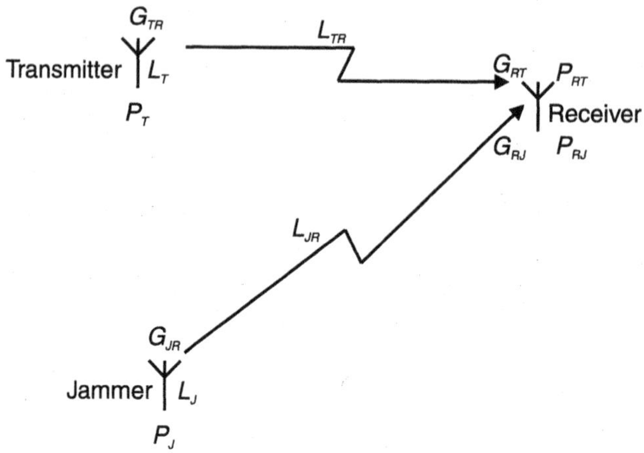

Figure 5.2 Factors impacting on power received from jammer and transmitter.

jammer: losses in the feeder system from the receiver's antenna to the receiver, and the receiver sensitivity.

For jamming to be effective, two conditions must be met: the power received from the jammer must be greater than the sensitivity of the receiver, and the JSR must be sufficiently large.

The value of JSR required to achieve a particular effect depends on the type of modulation used. A rule of thumb for FM voice and for data systems not protected by spread-spectrum EP techniques is that a JSR of 1 will lead to significant degradation of performance and a JSR greater than 2 will lead to an almost total loss of performance.

5.2.2 Jamming Signals

A jamming signal is created by modulating a baseband signal to the channel to be jammed. The most common types of jamming signals are frequency-modulated noise, amplitude-modulated noise, and CW.

Frequency-modulated noise can be used to jam both amplitude-modulated and frequency-modulated transmissions. The bandwidth of the jamming signal is controlled by a combination of the bandwidth of the baseband signal and the modulation index of the frequency modulator. A sufficiently strong frequency-modulated-noise jamming signal will capture a PSK, FSK, or FM receiver, preventing any of the original signal from being received. A weaker frequency-modulated-noise jamming signal will increase the bit error rate in data systems and slightly increase the noise in FM-voice systems.

The frequency-modulated-noise signal will cause additive noise in an AM receiver, causing significant signal degradation levels.

Amplitude-modulated noise can be used effectively as a jamming signal only with AM transmissions. Amplitude-modulated noise has little impact on FM transmissions, unless the noise levels are very large. The impact of an amplitude-modulated-noise jamming signal on an AM transmission will be to increase the additive noise in the receiver.

CW jamming involves transmitting only a carrier signal. This jamming can be effective against frequency-modulated transmissions via the capture effect and against AM transmissions when its signal level is much larger than the intended transmission. The drawback of CW jamming is its predictability, which makes it more easily overcome by EP techniques.

One advantage of using noise as a jamming signal is that it is inherently unpredictable. This increases the difficulty for EP techniques attempting to reduce the impact of the jamming, whether they are in the transmitted signal or in the receiver.

5.2.3 Types of Communications Jamming

There are a number of different types of communications jamming that may be employed. The main types and their characteristics are outlined in Figure 5.3.

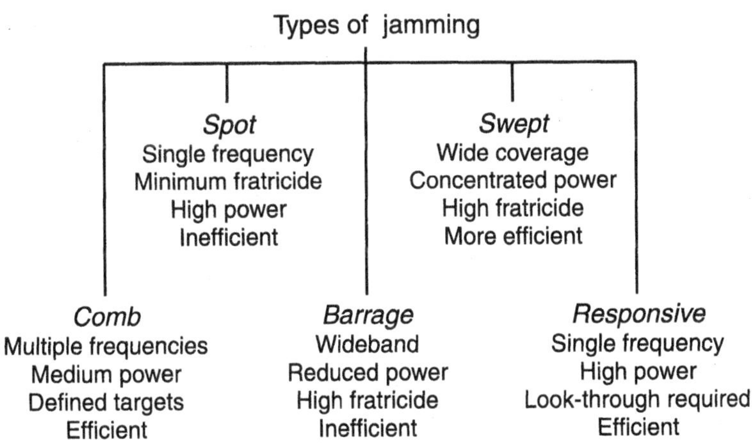

Figure 5.3 Types of communications jamming.

5.2.3.1 Spot Jamming

Spot jamming occurs on a single channel. As illustrated in Figure 5.4, a spot jammer consists of a generator for the jamming signal; a transmitter that translates the jamming signal from the baseband to the channel to be jammed and is likely to differ from a communications transmitter only in its higher output power; and an antenna that would usually be a directional antenna.

The generator for the jamming signal produces a baseband signal. As discussed previously, this would usually be noise. The transmitter modulates the baseband signal, placing it in the channel to be jammed. This modulation may be FM or AM. The bandwidth of the transmitted signal would usually be the same as the channel bandwidth of the system being jammed: 3 kHz for HF CNR and 25 or 50 kHz for VHF CNR. The transmitter also contains a power amplifier.

The directional antenna would usually be pointed at the receiver to be jammed. The use of a directional antenna both maximizes the jamming power at the receiver being jammed and minimizes the jamming power transmitted in other directions that might impact on friendly communications.

The spot jammer has the advantage that all of the jamming power is concentrated into a single channel, which will maximize the impact of the jamming on a receiver tuned to that channel. Jamming on a single channel also minimizes fratricide, since friendly communications systems using other channels are unlikely to be affected by the jamming unless they are very close to the jammer.

The disadvantages of a spot jammer are its lack of flexibility, the level of control required, and the lack of an in-built method to provide steerage.

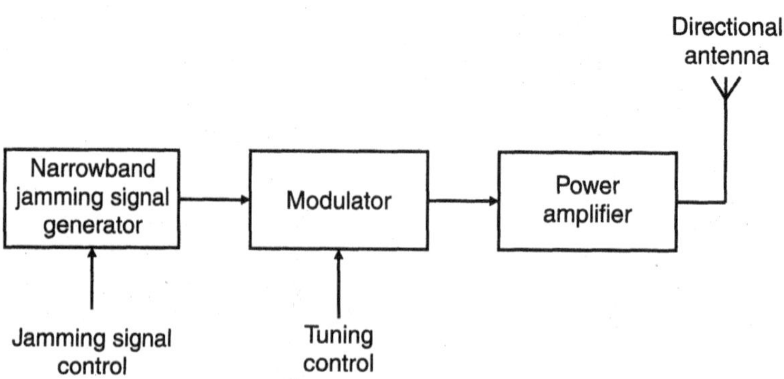

Figure 5.4 Outline structure of a spot jammer.

The spot jammer is tuned to a single channel, as would be done for a communications transmitter. It cannot rapidly change the channel being jammed. When using a spot jammer, jamming may have to be periodically suspended for short periods to allow friendly ES to evaluate the effectiveness of the jamming. This requirement may be overcome if the supporting ES receiver is located remotely from the jammer, so that its input is not swamped by the high-power jamming signal.

5.2.3.2 Barrage Jamming

In barrage jamming, the jamming signal is spread across a large number of adjacent channels. Because it jams a large number of channels, a barrage jammer requires relatively little information about the frequencies being used by the targeted system. Its jamming power is spread across a number of channels, reducing the jamming power in any one channel compared to a spot jammer, thereby reducing the impact of the jamming.

A barrage jammer (Figure 5.5) consists of a generator for the jamming signal; a transmitter, providing modulation and power amplification; and a directional antenna.

The generator for the jamming signal produces a baseband signal, which would usually be noise for a barrage jammer. A means would usually be provided to adjust the bandwidth of the transmitted jamming signal. This may involve control of the modulation process or the bandwidth of the baseband jamming signal.

The use of a directional antenna with a barrage jammer to prevent fratricide is even more important than with a spot jammer. The power output in each channel is lower than for a spot jammer, reducing the impact on adversary communications. Since a large number of channels are being

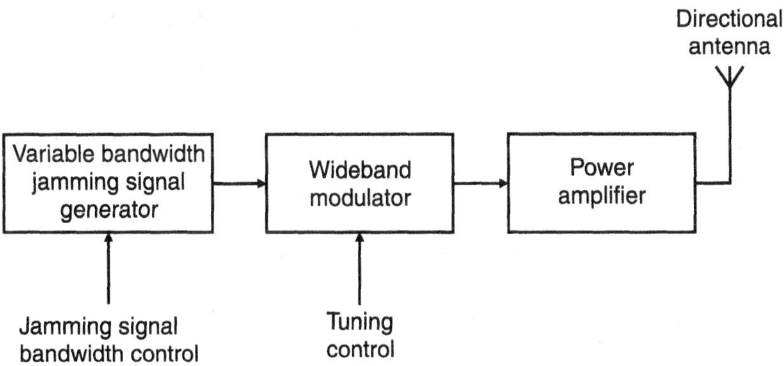

Figure 5.5 Outline structure of a barrage jammer.

jammed, the likely impact on friendly communications is higher. The use of a directional antenna will provide some protection to friendly communications by reducing the jamming power transmitted in other directions.

Evaluating the effectiveness of barrage jamming may require a periodic suspension of jamming to allow friendly ES to operate. A barrage jammer is easier to build than a communications transmitter with the same bandwidth and power because the quality of the transmitted signal is not important.

5.2.3.3 Swept Jamming

A swept jammer consists of a baseband source for the jamming signal, a transmitter whose output frequency is capable of being swept across a specified band, and an antenna, as shown in Figure 5.6. Like a barrage jammer, a swept jammer operates on a number of adjacent channels, and the use of a directional antenna with a swept jammer is important to minimize fratricide.

It may sometimes be possible to evaluate the impact of swept jamming on the adversary's communications system without suspending jamming. Because the swept jammer jams only one channel at a time, friendly ES can monitor the channels during the period that they are not being jammed. This requires close coordination between EA and ES equipment. The difficulty of coordination is likely to increase as the sweep rate increases.

A swept jammer is likely to have increased effectiveness over a barrage jammer because it concentrates all of the transmitted power in one channel, sweeping the frequency of this channel across a band. The impact of swept jamming is likely to be greatest on encrypted nets, where the jamming may cause the loss of synchronization in decryption systems. Even if jamming is stopped, further transmission of data cannot take place until synchronization has been reacquired.

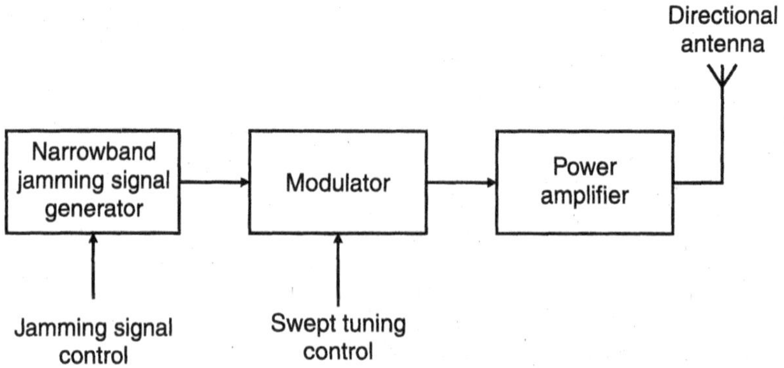

Figure 5.6 Outline structure of a swept jammer.

5.2.3.4 Comb Jamming

A comb jammer allows several channels to be jammed. These channels may be individually selectable by the operator, or may have a fixed spacing. The jammer transmits simultaneously on all channels being jammed.

Figure 5.7 shows the outline structure of a comb jammer. The comb generator generates the multiple carriers onto which the output of the jamming signal generator is modulated. The use of a directional antenna minimizes interference to friendly communications caused by the comb jammer.

The comb jammer has higher flexibility than the spot jammer because it allows more than one channel to be jammed. The power applied to each of these channels is reduced because the available power has to be shared between a number of channels. However, the efficiency is higher than for a barrage jammer, as the comb jammer jams selected channels rather than jamming a whole band.

Evaluating the effectiveness of comb jamming may require a periodic suspension of jamming to allow friendly ES to operate.

5.2.3.5 Responsive Jamming

A responsive jammer is a spot jammer (operates on a single channel) that jams only when it detects a transmission. A responsive jammer consists of a search receiver, a generator for the jamming signal, a transmitter to modulate the jamming signal, an antenna, and a control unit (see Figure 5.8).

The search receiver is used to detect transmissions that are to be jammed. The frequencies to be jammed are programmed into the control unit. When a transmission is detected, the control unit turns on the jamming for a

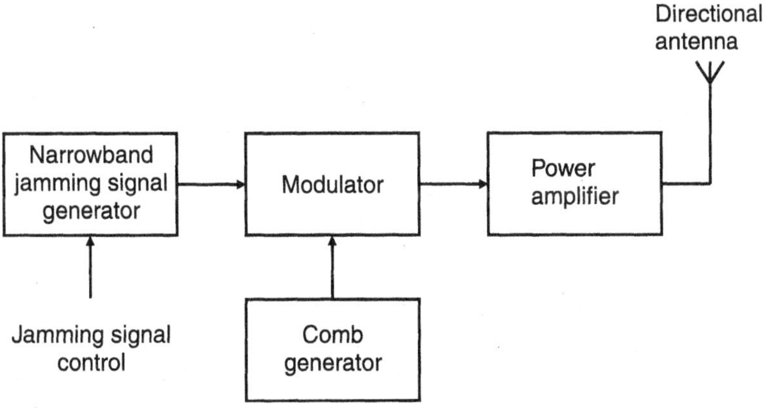

Figure 5.7 Outline structure of a comb jammer.

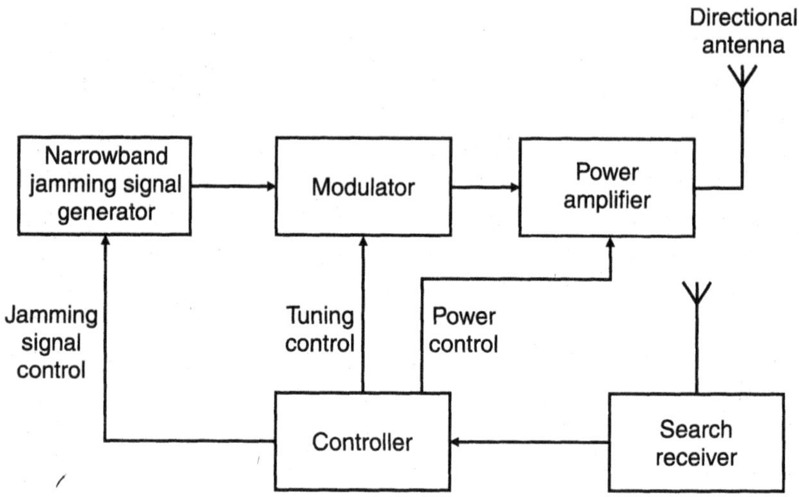

Figure 5.8 Outline structure of a responsive jammer.

specified period. Periodically, the control unit will turn off the jamming for a short look-through period, during which the search receiver will determine whether the transmitter is still active. If the transmitter is still active, the jamming will be recommenced. The look-through period may be of the order of 40 ms.

The antenna used by a responsive jammer may be directional, where the location of the receiver being jammed is both fixed and known, or omnidirectional.

A responsive jammer has all of the advantages of a spot jammer. It has the additional advantage of flexibility since its control unit can often be programmed with a number of frequencies to be jammed. By only transmitting when there is a signal to be jammed, the responsive jammer reduces its power consumption, reducing the weight of batteries required in portable applications; and reduces its electromagnetic signature, increasing the difficulty for an adversary to find and neutralize the jammer.

A responsive jammer may include facilities that allow it to operate as part of an automatic network of jammers in which only one jammer will be active on a particular channel at a time. This further reduces the signature of individual jammers.

A typical, modern, responsive jammer will allow a prioritized list of target frequencies to be entered. The jammer will then scan these frequencies, starting with the highest priority, until a transmission is detected. This transmission will then be jammed for a period. During a short period of look-through, the jammer will rescan its prioritized frequency list. If a higher-priority transmission is detected, jamming will switch to this channel.

5.2.4 Operational Factors

It is essential to note that, without careful planning and control of jamming resources, there is an inherent risk of fratricide to friendly communications, surveillance, and intercept assets. However, in order to be effective, jamming must be comprehensive. For instance, all HF, VHF, and radio relay systems of a targeted formation must be attacked simultaneously if communications are to be denied completely to an adversary commander. There are several operational factors that must be considered before a decision to jam is initiated:

- *Intelligence.* There may often be less value to be gained from denying the adversary use of the EM spectrum than there is in monitoring it to gather intelligence.
- *Alternatives.* EA is one weapon system available. Other forms of attack, such as artillery or fighter ground attack, may be more effective. As with all weapon systems, the target effect must be carefully considered before the method of attack is selected.
- *Security.* Jamming may announce to adversaries that their use of the spectrum has been compromised, prompting them to change frequencies and have to be found again. In certain sensitive cases, the ability to intercept and monitor adversary communications is so critical that jamming may never be considered for fear of compromising the capability. Jamming may also disclose future intention to act, such as a precursor to an attack.
- *Timing.* Like fire support, jamming must be timed to occur at the appropriate point in battle.
- *Selectivity.* EW assets are scarce and need to be used effectively. They are therefore controlled at the highest level and have associated comprehensive doctrines and procedures. Planning issues for EW are discussed in Chapter 6.

5.2.5 Standoff Versus Unattended Jammers

Unattended jammers (UAJ) are lightweight, low-powered jammers that are deployed in the near vicinity of a radio receiver. The basic principle is that a low-powered jammer close to a radio receiver will be able to jam far more effectively than a high-powered jammer farther away. In the example shown in Figure 5.9, a 10-W FM-voice transmitter (Tx) is 5 km away from its intended receiver (Rx). There is a UAJ about 1 km from the receiver and a standoff jammer (SOJ) 10 km away. To achieve a jamming power equal to the signal power, the UAJ is required to transmit at 400 mW, while the SOJ is required to transmit at 40W.

In addition, it is possible that at a distance of 10 km, the receiver may be screened from the SOJ, thus reducing the effect even further. A UAJ, however, is more likely to have line-of-sight to the victim receiver. In order to be effective, UAJs must be located close to the victim radio. Therefore, deployment plays a key role in the utilization of these types of devices. There are a number of ways in which UAJs can be deployed: withdrawing forces, special forces, rocket/shell, aircraft, or UAV.

UAJs deployed by withdrawing troops could be used to form an electronic minefield. An area seeded with UAJs would effectively jam the communications of advancing units, thereby creating confusion and delay. Employed in conjunction with conventional obstacles, such an electronic minefield might be used to prevent an adversary force encountering the obstacle from reporting its presence. Furthermore, unlike conventional antipersonnel

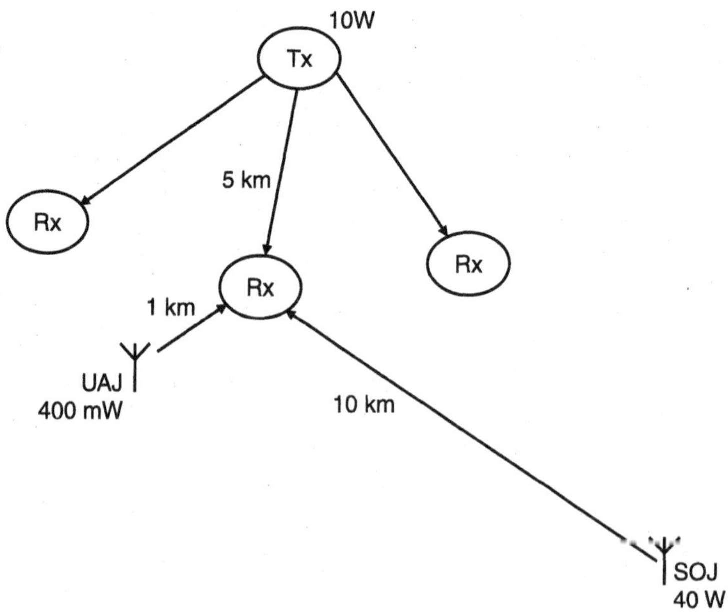

Figure 5.9 The efficiency of UAJ versus SOJ.

mines, UAJs do not injure civilians and can therefore be used in peacekeeping or peace support operations. The use of responsive jammers in this application, possibly also capable of remote control, would optimize both the jamming characteristics of the UAJ and its power utilization. UAJs operating as an electronic minefield would usually operate as a network so that only one jammer transmits on a channel at a time. This conserves power and reduces the signature of individual UAJs, making them harder for an adversary to locate and neutralize.

5.2.6 Jamming and the Tactical Communications System

Jamming has different levels of impact on the subsystems of the tactical communications system, which depend on the likely distances between jammer and target and the use of EP techniques in the target. The issues for jamming the tactical communications system are summarized in Table 5.1. Because effective jamming requires steerage from ES, the ability to carry out effective jamming of the tactical communications system is reliant not only on the issues directly associated with jamming that are discussed in this section but on the performance of ES, as discussed in Chapter 4.

5.2.6.1 Tactical Trunk Subsystem

Jamming of the trunk subsystem is more difficult than for the CNR subsystem because the elements of the trunk network are typically not deployed as far

Table 5.1
Jamming of the Tactical Communications System

Tactical Communications Subsystem	Vulnerabilities	Protection
Trunk	High antennas.	Directional antennas, long distance between receiver and jammer.
CNR	Omnidirectional antennas, short distance.	FH (sometimes), terrain screening.
Tactical data distribution	Omnidirectional antennas, short distance between receiver and jammer.	Heavy use of EP, including spread spectrum.
Airborne	Receivers on uplinks are exposed.	Receivers on downlinks may be protected from ground-based jammers.

forward as the CNR subsystem, although this is less so in low-level operations. The use of directional antennas in a terrestrial radio relay network also makes it difficult to achieve sufficient JSR for effective jamming. By using altitude to overcome the limitations of terrain, an airborne jammer lying on the axis of a terrestrial radio relay link may be able to jam communications in one direction. It is unlikely, however, that an airborne jammer could be sited so as to jam both directions on a duplex terrestrial radio relay link, as the only location from which this would be possible is directly between the two antennas. The high level of meshing that is usual in the terrestrial radio relay network means that jamming of a single node has little impact on the network unless that node is a choke point.

SCRA is more vulnerable to jamming than terrestrial radio relay because of its use of omnidirectional antennas and the smaller distance between some SCRA centrals and the jammer. The vulnerability of SCRA uplinks is increased by the placement of antennas at SCRA centrals on masts, which is required so that they can provide area coverage. Some protection is afforded to SCRA mobile terminals by the use of low antennas.

Trunk satellite links may be able to be jammed using tactical assets. In practice, such jamming is likely to be controlled at the strategic rather than the tactical level.

5.2.6.2 CNR Subsystem

A high proportion of tactical jamming is aimed at the CNR subsystem due to the employment of CNR in forward areas and its use of omnidirectional antennas. Jamming of CNR may target all of the stations on a net, all of the stations in a particular area, or one particular station. It may be used to deny communications or simply to force an encrypted net to operate in a nonsecure mode so that friendly ES can obtain information. Jamming of CNR can be hindered by the use of EP, especially FH.

5.2.6.3 Tactical Data Distribution Subsystem

Jamming of the tactical data distribution subsystem is aided by that subsystem's use of omnidirectional antennas and deployment in forward areas. The heavy use of EP techniques, especially spread-spectrum communications, will often make jamming of the subsystem difficult.

5.2.6.4 Airborne Subsystem

Receivers carried by the airborne subsystem are likely to be vulnerable to jamming due to their high position, which makes the employment of terrain screening difficult. This vulnerability is likely to be increased by the use of omnidirectional antennas, even for links that might employ directional antennas in their ground terminals. Antennas on ground terminals, however, may have a reduced vulnerability to jamming because the use of the airborne rebroadcast platform should facilitate the use of terrain screening.

5.2.7 Jamming Platforms

Jammers are deployed in a number of different ways, including mounted in ground vehicles and aircraft, man-portable, and artillery-delivered. The various platforms each impose constraints of weight, volume, and mobility, which combine to limit the capability of the systems.

5.2.7.1 Ground Vehicle

Deployment of a jammer in a ground vehicle, either a truck or armored vehicle, gives the jammer limited mobility on the battlefield. A ground vehicle carrying jammers is usually required to have the same or higher mobility as the force that it supports, for example, a jammer supporting a mechanized formation would usually be mounted in an armored vehicle.

Like all ground-based transmitters, the useful range of a ground-based jammer is limited by the terrain, especially at VHF and higher frequencies. In some circumstances, this limitation can be used to advantage, such as when a particular receiver on a net is being targeted and terrain screening is used to ensure that the remainder of the net is not aware of the jammer's effect.

Some systems allow jamming while the vehicle is moving; others require the vehicle to stop and a mast is deployed. The advantages gained from the use of a mast are reduced propagation loss at VHF and higher frequencies due to the increased height of the antenna, and therefore increased range.

A typical communications jammer would be capable of jamming at least one military band, for example, the HF band from 2 to 30 MHz or the VHF band from 30 to 88 MHz. More recent systems are capable of jamming over larger bands, sometimes as much as 100 kHz to 1 GHz. The jammer may have the capability to transmit simultaneously on a number of channels, making it possible to jam several targets at one time. The maximum, total, output power of a vehicle-transported jammer is likely to be at least 1 kW, and may be as much as 10 kW. A typical vehicle-mounted jammer is capable of operating from either batteries or from an external generator. The jammer's batteries would usually be charged whenever the vehicle's engine is running.

The jammer may have an in-built ES receiver to provide steerage, in the same way that a forward observer provides steerage for artillery. This receiver will operate over at least the frequency range of the jammer's transmitter and may be a narrowband receiver (scanner) or a wideband receiver. If the jammer is capable of look-through, it would be normal for the ES receiver to be able to determine the status of adversary transmissions during this time.

5.2.7.2 Man-Portable

Lightweight, portable jammers can be carried by dismounted soldiers, including special forces. Such jammers usually sacrifice transmitter power for weight, and are therefore designed to be deployed close to the receiver being jammed. They will usually also be operated with low antenna heights, which will further increase losses between jammer and receiver, and reduce their suitability for standoff jamming applications. The weight and bulk of a mast usually preclude its use with a man-portable jammer. Trees and other naturally occurring features may be used to provide additional elevation for antennas when the tactical situation permits.

A modern man-portable communications jammer is likely to cover similar portions of the electromagnetic spectrum to a vehicle-mounted jammer. It may also be capable of simultaneous transmission on a number of channels. The total output power of the man-portable jammer may be up to approximately 20W, or 100W using an amplifier appliqué. The total weight of the system is likely to be 10 to 15 kg. The operating time on the in-built batteries is likely to be a few hours. The man-portable jammer may have an in-built ES receiver, which is most likely to be a narrowband receiver.

5.2.7.3 Airborne

An airborne jammer may be deployed in a UAV, a helicopter, or a fixed-wing manned aircraft. Care must be taken in the installation of high-power jammers in aircraft to ensure that there is no interference with the operation of navigation equipment.

The main advantage is that, due to their elevation, airborne jammers overcome the limitations of terrain that reduce the performance of ground-based jammers. However, in doing so, they lose the flexibility of using terrain to jam a particular receiver while leaving the remaining receivers on the net unaffected.

A modern airborne jammer deployed in a medium fixed-wing aircraft is likely to be able to cover the spectrum from below the HF band up to 2 GHz. It may be capable of targeting both communications and noncommunications emissions. The output power is likely to be at least 1 kW. The jammer will take its power from the aircraft, and will be capable of operation as long as the aircraft remains airborne. The system will almost certainly be capable of jamming multiple channels simultaneously.

A jammer carried by a small, tactical UAV is likely to have similar properties to a man-portable jammer. Its capability is significantly enhanced by its greater operating height. It is unlikely that a UAV-mounted jammer will take its power from the aircraft, relying instead on batteries.

Jammers carried by other types of aircraft are likely to have intermediate capabilities. The major limitation in smaller aircraft is likely to be weight, which limits the output power of the jammer.

An airborne platform is likely to house both EA and ES assets, facilitating coordination between the different EW subdivisions. The in-built ES would normally be able to provide steerage for jammers carried by the platform.

5.7.2.4 Artillery-Delivered

An unattended jammer may be delivered by artillery. This has the advantage of allowing placement of the jammer in areas to which access cannot otherwise be gained. Such jammers are limited in both weight and volume. This delivery mechanism will place considerable physical stress on the jammer's electronics, both at the time of launch and on impact with the ground. A grenade launcher could potentially provide a similar delivery mechanism.

An artillery-delivered jammer would be designed to operate as an unattended jammer. It is most likely to operate as a responsive jammer, operating either when signals are received in preprogrammed channels or when any signal above a specified power level is received. A number of artillery-delivered jammers may be deployed as an electronic minefield, which may be used to deny communications to an adversary within a defined area.

Operation of an artillery-delivered jammer as a responsive jammer is advantageous both to reduce power consumption and thereby increase battery life, and to increase the difficulty for an adversary to find and remove the jammer. To conserve power, some jammers may be remotely activated by an HF command link. Due to constraints of weight and size, an artillery-delivered jammer is likely to operate in a single band (e.g., VHF CNR 30–88 MHz). The operating time is likely to be not much more than one hour of transmitting. It is unlikely that replenishment of batteries would normally be feasible for an artillery-delivered jammer.

5.3 Electronic Deception

The aim of electronic deception is to mislead or confuse an adversary. Like other forms of EA, electronic deception is not practiced in isolation; it forms part of the deception plan, which in turn is part of the commander's overall plan.

The main types of electronic deception are modification of the signature of friendly forces by changing the electronic profile (both communications and noncommunications emitters) of friendly forces—known as *manipulative electronic deception;* the use of dummy traffic, dummy communications, and electronic systems to create a representation of a force at the wrong location or with different capabilities—known as *simulative electronic deception;* and entry into adversary communications systems for the purpose of misleading and gaining information—known as *imitative electronic deception.*

Manipulative electronic deception and simulative electronic deception target adversaries through their ES systems, whereas imitative electronic deception targets adversaries through their communications systems.

Electronic deception is traditionally associated mostly with the CNR subsystem. The trunk subsystem tends to be difficult to target because of its distance from sites available for deception transmitters, its use of directional antennas, and the use of bulk encryption on all links. In the future, the commercial communications systems used as overlaid communications systems may provide significant opportunities for electronic deception due to their lack of in-built EP.

5.3.1 Manipulative Electronic Deception

Manipulative electronic deception, also known simply as manipulation, seeks to counter hostile ES and SIGINT by changing the electronic profile of friendly forces. Manipulation may use communications or noncommunications signals. This might be done by the deliberate changing of the technical characteristics of transmitters, controlled breaches of security, the transmission of additional traffic to creating a misleading impression of future intentions, or the use of alternate communications means or otherwise reducing traffic flows to alter the electronic signature associated with a particular operation such as preparation for an attack.

Successful manipulation requires a complete understanding of the characteristics of friendly transmissions and the relationship between variations in these characteristics and the various phases of an operation. Ironically, manipulation is likely to be more successful against a sophisticated adversary who has the ability to classify friendly transmissions and less successful against an adversary who does not have this capability.

Manipulation, used effectively, can cause dislocation of an adversary's intelligence efforts, resulting in the misdirection of ES and other intelligence assets to unproductive tasks. Without adequate steerage, it is then difficult for an adversary to undertake effective EA. This can reduce the impact of both adversary ES and EA on friendly communications.

5.3.2 Simulative Electronic Deception

Simulative electronic deception, or simulation, uses dummy electromagnetic emissions to create a representation of a force in the wrong location or to imply capabilities different from those actually possessed by the force. These emissions may come from either communications or noncommunications emitters.

There are three subdivisions of simulation: unit, system, and activity.

- *Unit simulation* is the use of actual equipment or simulators to make it appear that a unit is in a particular location.
- *System simulation* is the use of systems whose signature indicates a particular type of organization. For example, the use of elements of the trunk subsystem usually indicates the presence of a brigade or larger force in an area.
- *Activity simulation* is use of communications or noncommunications emitters in such a way as to simulate a force with a particular intention or engaging in a particular activity.

5.3.3 Imitative Electronic Deception

In this form of deception, EW operators attempt to enter adversary nets and imitate a station to mislead or gain information. Another form of intrusion is the retransmission of previously recorded, genuine net traffic from one or more of the adversary's nets. This technique has the greatest potential to create confusion when played at crucial times in attack or defense.

Imitation may take a number of forms, including *nuisance intrusion,* which involves the transmission of voice messages or data to distract operators or to confuse data communications equipment into thinking that a channel carrying traffic is congested; *planned-message intrusion,* designed to mislead an adversary or give false instructions; *cryptographic intrusion,* in which messages are inserted into an encrypted system; or *deception jamming,* which is normally applicable to radar systems, and in which signals are transmitted to the radar to indicate targets in false locations.

Imitation requires equipment capable of convincingly duplicating the operation of adversary equipment. Successful imitation requires a transmitter compatible with the intended receiving equipment, including frequency band, channel boundaries, and modulation, preferably using captured adversary equipment to ensure that the external characteristics of the signal are authentic; a suitable transmitter location such that the deception emissions will be received by the intended receivers, given the available transmitter power; and a proficient linguist (if voice transmissions are used) or operator capable of imitating the adversary Morse style (if CW is used).

Imitation is particularly difficult in fully encrypted nets. In this case, the only feasible options are likely to be to use captured adversary equipment before the key settings can be changed or to rebroadcast old traffic in an attempt to create confusion. Once again, old traffic will only be useable for the period between its intercept and the next time the encryption key is changed.

The use of imitation in an adversary's trunk subsystem is made difficult by the use of bulk encryption on trunk links, the use of directional antennas, and the distance between the elements of the trunk network and locations in which a deception transmitter can be placed.

While the use of EP, including spread-spectrum techniques and encryption, in the tactical data distribution subsystem hinders imitative deception, recording transmissions and replaying them at high power levels may in some circumstances allow a false tactical picture to be inserted into an adversary's systems.

5.3.4 Example Electronic Deception Techniques

Examples of the use of electronic deception techniques include transmitting dummy traffic to simulate the presence of a unit or asset that has moved to another location under radio silence; using trunk communications within a battalion area; indicating the presence of a larger force; leaving in place a significant part of a force's communications system while the force moves to another location; transmitting messages with misleading content in plaintext or using easily broken codes; and adding dummy messages to a link to indicate the buildup to a significant activity such as an attack.

Before the Battle of Midway in World War II, U.S. forces knew that a major Japanese offensive was planned, with the two most likely locations for an attack being Midway Island and the Aleutian Islands [2]. From intercepted Japanese radio messages, it was also known that an attack was planned at a location whose code name was AF. In order to confirm the location, a message was sent from Midway in a low-grade code, saying that the island was short of water. Shortly after this, a Japanese message was intercepted saying that AF was short of water, confirming that AF was, in fact, Midway. Based on this information, the U.S. forces were able to concentrate in the vicinity of Midway and defeat the Japanese attack.

The deception operation surrounding the move of the First (U.K.) Armoured Division during the Gulf War is a more recent example, where a division's presence was simulated by the broadcast of previously recorded radio traffic while the division was moving several hundred kilometers to the west under radio silence. It should be noted that signature equipment had to remain in place and on the air. In addition, a close liaison with U.S. forces was necessary, and, obviously, strict radio silence by the moving forces had to be maintained. Electronic deception is rarely effective on its own, and must be considered as part of a complete deception plan. Dummy equipment was erected to simulate troop concentrations and representative smoke and noise signatures also had to be considered. The activities of the media were taken into account and certain trusted members accompanied the main force as it moved west. In addition, a conscious decision had to be taken to allow SIGINT units of the Iraqi army to operate almost unhindered so that incorrect EOB data reached the Iraqi commanders.

5.3.5 Received Power

Like jamming, electronic deception is aimed at an adversary's receiver receivers. To be effective, deception must deliver sufficient power to the inputs of these receivers so that they can monitor the transmission. Unlike jamming, however, electronic deception does not rely on delivering high powers to adversary receivers, because deception transmitters are not competing directly with an adversary's communications transmitters.

Electronic deception may target all the receivers associated with a particular part of the adversary's communications system (e.g., a CNR net or group of ES stations), or it may target a single receiver. Like jamming, electronic deception requires ES support to assess its effectiveness.

The power received by the receiver's antenna from the deception transmitter depends on the power generated by the deception transmitter's power amplifier P_T, the loss in the feeder system to the transmit antenna L_T, the gain of the transmit antenna in the direction of the receiver G_T, the propagation loss due to the path between the deception transmitter and receiver L_{TR}, the loss in the feeder system to the receive antenna L_R, and the gain of the receiver's antenna in the direction of the transmitter G_R.

With the losses and gains expressed as power ratios, the power received from the transmitter P_R can be expressed as:

$$P_R = \frac{P_T \times G_T \times G_R}{L_T \times L_{TR} \times L_R}$$

These factors are illustrated in Figure 5.10.

Sufficient power is available at the receiver for the monitoring of the transmission if the received power P_R is greater than the receiver sensitivity P_{Rsens}, that is,

$$P_R \geq P_{Rsens}$$

5.4 Neutralization

Neutralization is the term used to describe the use of high levels of electromagnetic energy to disrupt the operation of, or to permanently damage, electronic equipment. The power levels incident on a target required to achieve these effects are very much higher than those required for jamming, which are higher again than those required for deception and communications. All electronic equipment, not just communications systems, and even some nonelectronic systems are potentially vulnerable to these neutralization weapons.

In the past, neutralization has focused on the effects of electromagnetic pulse (EMP); however, other neutralization weapons based on nonnuclear high-power RF technology are being investigated and have the ability to damage, dazzle, interrupt, or even destroy. Lasers also have the potential to disrupt and damage a variety of systems. These systems are examined in Section 5.4.1, with further detail on a promising new area in RF weapons—RF directed energy weapons, which is discussed further in Chapter 7.

5.4.1 Laser Weapons

High-power lasers are capable of generating sufficient power to cause a range of disturbance and damage to a variety of targets, including optical or infrared sensors, and structures at short ranges [3].

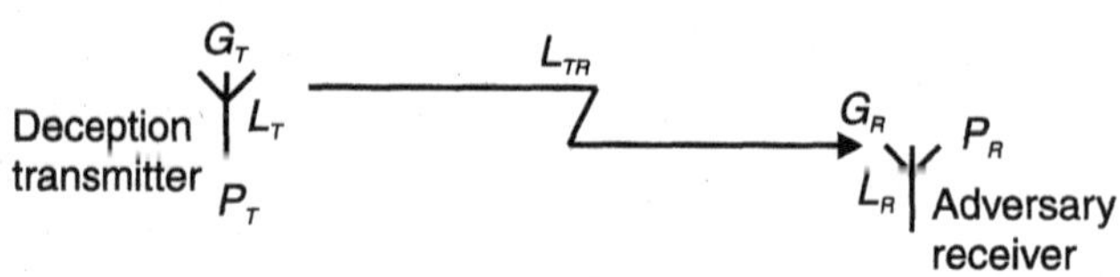

Figure 5.10 Factors affecting the power received from a deception transmitter.

5.4.1.1 Conventional Lasers

It is possible to generate sufficient power with conventional lasers to damage or confuse sensors at ranges up to 10 km, assuming that line-of-sight exists between laser and target. The potential vulnerability to such attacks is increasing with the increase in the use of electronic sensors on the battlefield.

Lasers have already found use for target designation and range-finding in a number of environments, and have immediate application in close-in self-defense systems to disrupt the operation of sensors on infrared homing missiles, which might be used against ground vehicles or aircraft. The same technology used by a missile could be used be used to dazzle or damage the sensors on a target that might enable the target to activate and guide its defense systems. International law prohibits the use of laser weapons that intentionally blind personnel [4]: "It is prohibited to employ laser weapons specifically designed, as their sole combat function or as one of their combat functions, to cause permanent blindness to unenhanced vision, that is to the naked eye or to the eye with corrective eyesight devices." The use of weapons that permanently blind personnel as a side effect of their main purpose is arguably permitted. While not part of communications EW, laser protection for personnel is an important aspect of defense against laser weapons.

In the future, it is likely that laser weapons will be used not only by conventional military forces but also by terrorists.

5.4.1.2 Higher-Energy Laser Weapons

Higher-energy laser weapons (HELWs) could feasibly be used as close-in point defense systems to destroy incoming missiles and smart bombs at an effective range of 10 km, given line-of-sight. In the land environment, with the impact of terrain and the tactics adopted by land forces to obtain protection against conventional threats, ranges are likely to be significantly shorter. This is primarily based on the U.S. research effort that has had some success, particularly at high altitudes. The demonstration of an airborne laser (ABL), mounted in a 747 aircraft and capable of destroying missiles, is scheduled for 2002. Work on the development of HELWs that provide capabilities suitable to land and sea environments is under way in a number of countries. While the surface deployment of HELWs faces greater problems as a result of atmospheric absorption and refraction, especially in environments subject to salt spray or high dust levels, such systems are likely to follow the development of the high-altitude ABL. The development of a tank-mounted CO_2 laser called HELEX [5] by the German government is an example of a surface-deployed HELW. Such weapons are designed to melt aluminum.

5.4.2 High-Power RF Weapons

It has been known for many years that the EMP generated by a nuclear explosion is capable of permanently damaging electronic circuits over a significant range and disrupting the operation of electronic circuits over even longer ranges. For this reason, one form of equipment ruggedization specified for electronic equipment is to provide protection against EMP and other sources of higher-power RF radiation. The levels of protection required are specified in a number of military standards [6]. Of particular concern are safety and arming circuits for weapons systems, for which the highest levels of protection are specified.

The development of nonnuclear RF weapons that can disrupt or damage communications and electronic equipment over tactically useful ranges presents significant challenges. The major reason for this is the dilution of transmitted power with distance from the transmitter, just as the size of ripples caused by a stone dropped into a pool of water decreases as the ripples move out from the point of impact. Notwithstanding these difficulties, their potential tactical value is great as a means of attacking equipment that is not vulnerable to jamming, such as computers, engine-management systems in vehicles, and navigation systems in aircraft. Many countries, including several NATO members and Russia, have programs under way to develop high-power RF weapons.

The issues associated with the development of high-power RF weapons are discussed in Chapter 7, along with possible tactics for their use and defense against such use.

Endnotes

[1] U.S. doctrine on EA can be found in:
U.S. Army Field Manual FM 6-20-10, "Tactics, Techniques, and Procedures for the Targeting Process," May 1996.
U.S. Army Field Manual FM 34-1, "Intelligence and Electronic Warfare Operations," September 1994.
U.S. Army Field Manual FM 34-45, "Tactics, Techniques, and Procedures for Electronic Attack," June 2000.

[2] Smith, M., *The Emperor's Codes: Bletchley Park and the Breaking of Japan's Secret Ciphers*, London: Bantam Press, 2000.

[3] See, for example, Anderberg and Walbarsht, *Laser Weapons: The Dawn of a New Military Age*, New York: Plenum Press, 1992.

[4] International Committee of the Red Cross, *International Law Concerning the Conduct of Hostilities,* Protocol on Blinding Laser Weapons (Protocol IV), 1996.

[5] Linder, F.W., "Laser Weapons for Tactical Operations," *MILTECH,* 1987, pp. 125–135.

[6] Standards specifying protection for military equipment against high-power RF signals include:
MIL-STD-464, *Electromagnetic Environmental Effects—Requirements for Systems,* U.S. Department of Defense, March 1997.
MIL-STD-188-125-1, *High-Altitude Electromagnetic Pulse (HEMP) Protection for Ground-Based C4I Facilities Performing Critical, Time-Urgent Missions Part 1 Fixed Facilities,* U.S. Department of Defense, July 1998.
MIL-STD-188-125-2, *High-Altitude Electromagnetic Pulse (HEMP) Protection for Ground-Based C4I Facilities Performing Critical, Time-Urgent Missions Part 1 Transportable Systems,* U.S. Department of Defense, March 1999.

6

Land EW Command and Control

6.1 Introduction

Chapters 3 to 5 have described the three EW functional areas of ES, EA, and EP and have demonstrated the technical characteristics of the various techniques available within each area. Like any other military system, the employment of EW also requires a planning and coordination process, which is the subject of this chapter. Like all other planning on the battlefield, EW planning does not exist in isolation and must form part of the broader planning process for an operation.

EW planning itself is inherently based on knowledge of the target equipment and systems that are primarily associated with an adversary's communication and information systems (CIS). EW is not alone in this regard. Friendly CIS planning should also take into account adversary EW, mobility planning, the adversary's countermobility capabilities, and so on. Furthermore, there is always a tension between attacking a target to neutralize or destroy it, and allowing it to operate under surveillance to obtain intelligence. Therefore, once the EW plan is complete, it must be tested against the CIS plan to ensure that they do not conflict.

EW planning is primarily about ES and EA, that is, the targeting of adversary systems. The employment of EP, as it is aimed at the protection of friendly capabilities, is normally considered to be part of the CIS plan. This chapter is primarily oriented toward ES and EA, as illustrated in Figure 6.1.

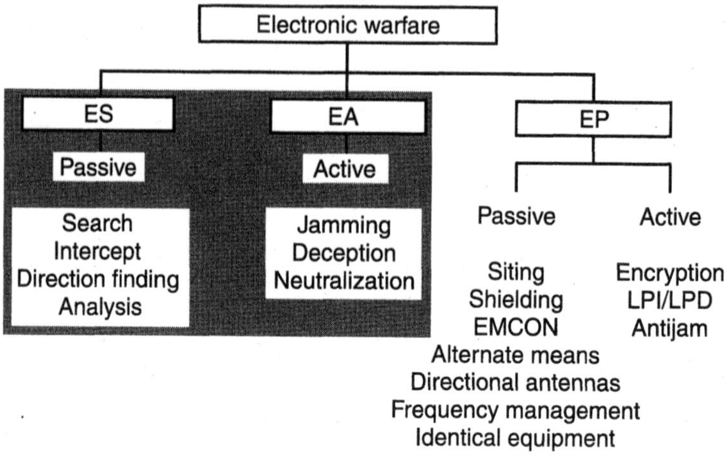

Figure 6.1 EW architecture.

The employment of EW is traditionally controlled at the highest level, as are a number of other key assets such as artillery. The reasons for this are scarcity of resources, security, and coordination.

Scarcity of Resources There are many more targets on the modern battlefield than there are resources to attack, or even just to detect and acquire, them. It is therefore crucial that resources are not wasted (e.g., by using two jammers against a single target) and that resources are allocated to targets with the highest payoff.

Security The use of an electronic asset always tends to gives away information on friendly capabilities. Like any other transmitter, a jammer's location can be identified by DF; arguably, given the higher transmit powers, a jammer is easier to DF than a communications transmitter. Even the use of passive means, such as ES, may give away a capability, which could arise from the physical observation of the ES systems or from the use of information obtained by ES in the planning and conduct of operations.

Coordination The EW plan needs to be coordinated across the force and with other plans. At the most basic level, coordination is required to resolve the basic tension on the battlefield between destroying a target and observing it for intelligence. This coordination is more than deconfliction, however, it is required to maximize the impact of all action taken. Coordination is used to ensure that one part of the force does not jam a transmission against which another part is using ES. Coordination is also essential in other areas, such as preventing artillery from destroying a target that is providing valuable intelligence. Examples of coordination required between EW and other battlefield planning include CIS planning, deception planning, intelligence planning, and artillery planning. In a joint operation, coordination with sea and air planning would also be required.

While EW is controlled at the highest level, EW assets may be provided in direct support of a part of the force, giving a lower-level commander control over their use for a given operation, in a particular area, for a particular time, and under specific rules of engagement. This approach is almost identical to that used for artillery and deception operations, taking into account scarcity of resources, security, and coordination while preserving mission command and thereby maximizing the flexibility of tactical commanders.

This chapter describes the activities that need to occur for EW planning. The terminology used is consistent with U.S. Army doctrine [1], but does not set out to describe in detail the EW planning process of any particular army. While the functional grouping of activities may vary, the basic processes that must be carried out in EW planning are the same.

6.2 EW Unit Organization

Before examining the tactical EW planning process, we will briefly discuss the considerations involved in structuring EW units and assigning assets to them.

As previously stated, EW is controlled at the highest level, with EW assets being provided to lower-level commands in support of a particular operation. EW units are often allocated as direct-command units at the corps or division level with elements deployed at the brigade and battalion levels. In conventional operations, a corps EW unit would be expected to provide ES and EA support to the corps across its frontage and as close to the adversary as possible.

The allocation of ES and EA assets (i.e., the dimensioning of the EW system) should take into account a number of criteria: the minimum levels required for operation, quality, operation on-the-move, the area coverage required, capacity, the electronic signature of transmitters, redundancy, platform mobility, command and control, and economy.

Minimum Levels for Operation Some EW techniques require more than one platform to provide even the most basic level of performance. DF, for example, cannot locate a target without at least two stations, although three or four are preferred.

Quality Providing more assets can increase the quality of information obtained from some ES systems, such as DF. Having reached a certain level, the quality may not improve significantly with the allocation of further assets.

Operation On-the-Move Some platforms, especially airborne platforms, enable operations to continue as the platform moves. Others, such as ground vehicles that may require antennas mounted on long masts and are subject to the effects of terrain, may only be able to operate while stationary. In the former case, continuous operation may be achieved by using a single platform; in the latter case, a minimum of two platforms is required to achieve continuous operation.

Area Coverage Required At VHF and higher frequencies, the coverage of EW systems is essentially limited to line-of-sight (i.e., the area covered is primarily limited by terrain). If a large area coverage is required, it may be necessary to subdivide this area, with independent equipment providing coverage of each subarea.

Capacity ES and EA systems are able to provide coverage of a certain number of channels. For example, a jammer may be able to operate on only one channel, or an intercept platform may be able to monitor 10 channels simultaneously. If higher capacities are required, some replication of equipment and personnel is normally required.

Electronic Signature In some types of operations, especially deception, the electronic signature may need to include transmissions from more than one location.

Redundancy Provision must be made for the continued effective operation of EW systems in the event of equipment failure due to attack by an adversary. This attack may be a physical attack, such as artillery, or an electronic attack, such as masking being used to jam an ES receiver. Redundancy is of particular importance for a system such as a jammer that advertises its location by transmitting at a high power level.

Platform Mobility Ground-based platforms may have low rates of movement. In order to allow for contingencies, it may be necessary to position assets in advance for a future operation while maintaining support for the current operation. This may require that more units be available to planners than would otherwise be required. Additionally, EW assets must be as mobile as the force being supported, and arguably more mobile if they are to be in position ahead of the force to support operations.

Command and Control Like any other part of the force it supports, EW requires an appropriate structure for command and control, including tasking and reporting. This limits the number of EW assets that can be deployed in a single unit. Command and control can also be problematic if EW assets are dispersed too thinly. Communications place a significant limit on EW command and control, and it is essential that the EW command facility can communicate to each of its EW assets.

Economy EW assets are scarce resources and should be used sparingly and controlled centrally. The collocation of EW elements in a single platform or at a single location is influenced by the requirement for similar or the same coverage from assets, platform capacity, and support requirements.

Requirement for Similar/Same Coverage When a search receiver passes information to an intercept receiver, the intercept receiver must be sited in a location from which it can receive the same signals as the search receiver. Search and intercept receivers are therefore naturally collocated on the same platform. A similar constraint applies to an ES receiver supporting EA. Although this means that the jammer and supporting intercept receiver are in the near vicinity of one another, they are normally on different platforms.

Platform Capacity Factors such as weight, volume, availability of power supply, and mutual interference may limit the number of EW assets that can be located on a given platform. This is likely to be a particular constraint on small air platforms, such as tactical UAVs.

Support Requirements For best results, ES should support EA. Minimizing communications requirements and providing similar coverage suggest collocation, but mutual interference suggests remote location. Additionally, support of EA is a secondary task for ES assets, which will normally be deployed for optimal coverage to achieve intelligence collection. EA assets may therefore be forced to rely on their look-through capability to assess jamming effectiveness.

As equipment becomes more integrated, providing complete ES/EA systems in one unit, the issues for collocation of assets are becoming less relevant to planning.

6.2.1 Electronic Attack

A minimum of one transmitter is required for EA. For ground systems, however, at least two transmitters are normally required, since operation on-the-move is not usually possible due to a combination of terrain and the physical characteristics of antennas. For airborne systems, a minimum of three platforms is probably required to allow for refueling and platform maintenance. Even if air-to-air refueling is available, it is unlikely that an EA system will be able to operate during refueling. Providing coverage for a corps area in conventional operations probably requires simultaneous operation of at least two separate ground or airborne EA platforms at VHF and higher frequencies, although this will depend on the terrain in which the systems are planned to operate. Adequate coverage may be addressed by allocation of a jamming capability to each division, consisting of a minimum of two ground-based jammers or three airborne jamming platforms. The capacity required is likely to be more than one channel per division, and can be addressed either by providing multichannel jammers or extra jamming transmitters.

This analysis suggests that a minimum of three airborne or two ground-based jammers per division is required, increasing to four airborne or three ground-based jammers per division if redundancy is required. However, on the modern battlefield, more jammers could always be used and the number allocated to ground forces is normally as a result of funding compromises in the provision of an adequate force structure.

Finally, EA must normally be supported by ES for basic steerage, in the same way that artillery uses a forward observer to adjust fire. This supporting ES may be integral to EA systems, or provided by separate equipment. The use of separate ES assets reduces the number of ES assets available for other ES tasks, and reduces the flexibility of deployment of these assets. Normally, a compromise is reached in which the EA asset has its own integral look-through capability, which is augmented where possible by separate ES assets.

Current U.S. doctrine allocates three airborne (helicopter) and three ground-vehicle-based jammers to each division [2]. An additional three ground-vehicle-based jammers are allocated to each corps. All have an integral ES capability (intercept and DF).

The above analysis is based on conventional, high-density operations. For dispersed operations, larger numbers of EA systems are likely to be required due to the range limitations imposed by terrain. This increase in numbers will be greater for ground platforms than for airborne platforms.

6.2.2 Electronic Support

A minimum of one search receiver is required, with operation on-the-move requiring at least two systems. Allowing for redundancy increases this to three systems. The provision of coverage to a corps area in most terrain would be possible with systems allocated on a divisional basis, leading to a requirement of three search facilities per division, with a total of nine per corps.

The same criteria apply to intercept, except that the number of channels that are required to be monitored simultaneously may dictate the use of a larger number of intercept systems. Because of the requirement for similar communications coverage, search and intercept facilities are likely to be collocated, even if the two services are provided by separate equipment. For example, a tactical intercept shelter may contain two to four intercept receivers, whose operators are directed by the product of a search receiver located in the same shelter.

For DF, a minimum of two stations is required, with three usually used to provide sufficient accuracy. For ground systems, twice this number is required to allow the baseline to move. The coverage of a corps area can probably be achieved in most terrain by the allocation of these units to each division, leading to a requirement for 6 units per division, or 18 units per corps. The use of airborne platforms may reduce these numbers to 4 per division, or 12 per corps.

Current U.S. doctrine allocates major communications ES assets [2] on the basis of 12 airborne ES platforms per corps, organized into two systems of 6 platforms, plus 6 ground-based (vehicle or man-pack) systems. Each division has 6 ground-based systems. An ES payload may also optionally be carried by a short-range, tactical UAV.

The search, intercept, and DF functions may be integrated into a single receiver, which is most likely to occur with a digital receiver. One important advantage obtained by doing this can be the speed of the handoff of targets from search to intercept and DF, which can enhance ES against short-term and rapidly changing signals.

In order to provide effective DF, ES systems tend to be deployed in groups of approximately three stations. Each of these groups would have its own internal communications, as well as a rear link to the command and processing facility. In some systems, equipment required for DF is separate from that required for search and intercept. This is particularly true for older equipment. In this case, all stations in a group may be capable of DF, but only one station may be capable of search and intercept.

This analysis is based on conventional, high-density operations. For dispersed operations, larger numbers of ES systems are likely to be required due to the range limitations imposed by terrain. This increase in numbers will be greater for ground platforms than for airborne platforms.

6.2.3 EW Command and Control

In addition to ES and EA assets, effective EW requires both an EW command system and an EW processing and dissemination system. Communications between ES, EA, and EW command, processing, and dissemination systems may be direct via encrypted, data-enabled CNR, or may be carried by another part of the tactical communications system. If passed across the trunk subsystem, EW communications are double encrypted (i.e., encrypted by the EW units and then encrypted again by the bulk encryption on the trunk links).

A typical arrangement of EW assets in a division is shown in Figure 6.2.

It is usual that the EW command facility and the EW processing and dissemination systems are collocated with the highest level headquarters being

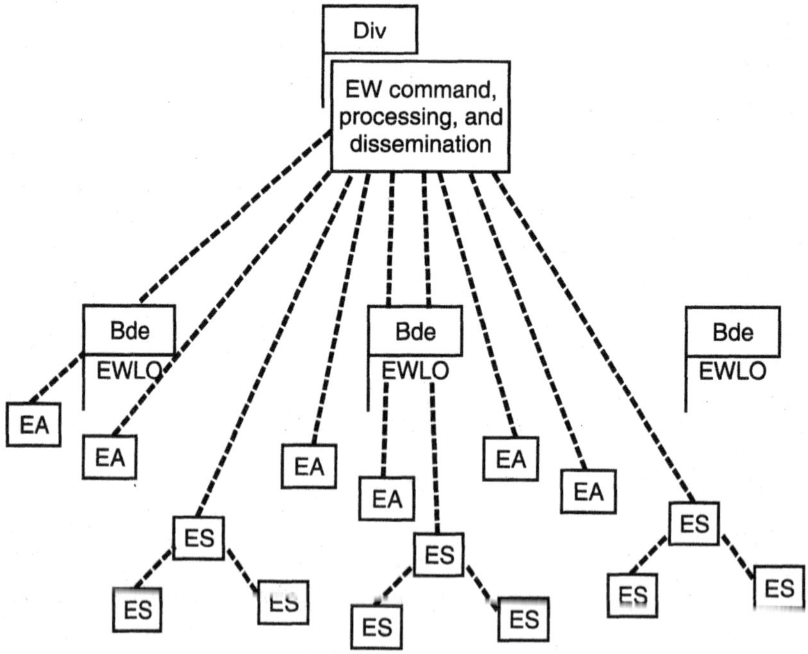

Figure 6.2 EW system.

supported. This requirement has existed traditionally because of the amount of information flowing between these elements and the scarcity of communications resources. In a future networked battlefield, this requirement may be relaxed by the availability of ubiquitous, high-capacity communications not only throughout the battlefield, but also in rear areas. EW liaison officers (EWLO) are typically stationed at lower-level supported headquarters, which is brigade in the case of the example in Figure 6.2. The processing and dissemination functions take data derived from ES and pass it into the broader intelligence processing system.

Unprocessed EW data is normally classified more highly than other operation data due mainly to a need to protect the existence and capability of EW assets. Therefore, EW product is rarely distributed directly, but is sanitized first and then collated with collateral intelligence information before distribution. Sometimes, some sensitive data is never declassified and is only distributed directly to commanders who are cleared to receive it.

6.3 The Tactical Planning Process

Most armies have a defined process by which operational plans are developed. These processes are all based on a planning sequence that begins with an analysis of the aim of the activity and the factors that have to be taken into account, and then focuses on the production of one or more candidate plans, from which the most suitable is chosen for execution. The U.S. process is known as *Intelligence Preparation of the Battlefield* (IPB) [3]. The United Kingdom's *Estimate* and the Australian *Military Appreciation Process* are other examples.

The sequence of activities in the IPB is shown in Figure 6.3.

Receipt of Mission The mission is defined by the commander, and will be based on orders received from the higher headquarters. The mission should not simply be a task, but should make clear the commander's intent (i.e., the mission should state—at least in broad terms—both what is to be achieved and why). How the mission is achieved is defined by the remainder of the planning process.

Mission Analysis Mission analysis takes the assigned mission and identifies tasks to be carried out to meet this mission. These tasks may be explicit (i.e., they are stated as part of the mission), or implicit (i.e., they are required to achieve the mission, but not explicitly stated as part of the mission).

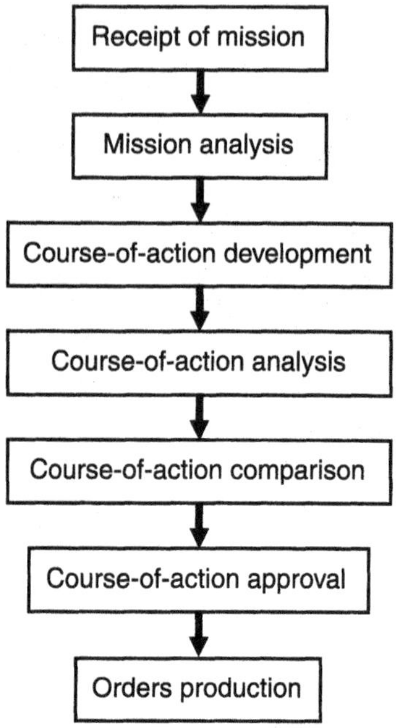

Figure 6.3 Sequence of actions in the IPB.

Implicit tasks are of particular importance in the gathering of information, such as the adversary electronic order of battle, that occurs on a continuing basis, not just in response to a specific tasking. As well as identifying what is to be achieved, mission analysis identifies constraints and limitations implied by the mission.

Course-of-Action Development A number of possible courses of action are identified to achieve the mission. At this stage, the details of the feasibility of the individual courses of action are not important.

Course-of-Action Analysis Each course of action is war-gamed. This step of the process takes into account the impact of adversary courses of action on each of the candidate-friendly courses of action. In a large staff, the war-gaming may be carried out with the operations staff taking the part of the friendly forces and the intelligence staff taking the part of the adversary. In a small staff, or where the process is carried out by a single person, war-gaming is still performed but without separate actors for the two sides. A course of action may be refined based on the outcome of the war-gaming.

Course-of-Action Comparison The results of the war-gaming permit a detailed comparison of the strengths and weaknesses of each candidate course of action. These are compared and one course is chosen, possibly with a small number of options.

Course-of-Action Approval The outcome of the course of action comparison is presented to the commander, who confirms the solution. If a number of options are presented, the commander selects the option to be implemented.

Orders Production The final step in the planning process is the production of orders. These will usually take the form of a main document that sets out the broad plan, with a number of annexes providing details of specific areas such as the fire plan, communications plan, engineer plan, and EW plan.

Regardless of how it is driven from above, EW planning itself consists of EA planning and ES planning, each of which is coordinated with related activities such as CIS planning and artillery planning.

6.4 The EW Targeting Process

EA is only one of the means available to a commander to degrade or destroy an adversary's combat capability. In this sense, EA can be regarded as one of the fires available to a commander, along with assets such as artillery and attack helicopters. Because EA is controlled at the highest level, the largest EA planning capability lies at this same level.

The EA planning process ensures that actions taken are consistent with the basic constraints imposed by the scarcity of resources, security, and coordination. A targeting process, such as that used to govern the employment of EA, may be divided into four parts:

- *Decision.* The first part of the targeting process provides its focus, a targeting plan, and input to the intelligence collection management process, including guidance on the prioritization of targets.
- *Detection.* Targets identified by the decision process must be detected through the use of a combination of ES and intelligence assets. Functions of detection may include finding operating frequencies or mapping network structures to identify a critical node or link for EA.
- *Delivery.* EA assets are used to attack targets identified by the decision process, using information provided by the detection process.

- *Assessment.* Any form of fire requires an observation of its effect. For EA, this observation is provided by ES. The final result of the assessment process is a recommendation for or against reattack.

The sequence of activities involved in the targeting process and the numbers of the sections in which they are discussed are shown in Figure 6.4.

While the following discussion is expressed in terms of the targeting process for a particular operation, functions such as decision and detection would usually be ongoing. This is required to build up an adequate database of the adversary EOB, and is a basic part of the intelligence process. The lead-up to a new operation will simply provide more specific and detailed guidance for these functions.

Sections 6.4.1 to 6.4.4 examine each of these parts of the targeting process. The greatest emphasis is contained in Section 6.4.1 on decision, which is the heart of the planning process.

6.4.1 Decision

Decision provides the EW planning and control functions of the targeting process. Decision takes its direction from the broader operational plan, combines this with information provided by the detection and assessment functions and from other sources, and tasks the available assets. The decision process is carried out by specialist EW planning staff, who would usually be collocated with the headquarters that controls EW.

The key outputs of the decision process are a high-payoff target list; guidance for collection of information to support the delivery of EA, mostly guidance for ES; and tasking for ES and EA assets, which may take the form of an EW annex to an operational order.

These outputs are generated prior to an operation. While the operation is carried out, the decision process continues, updating all its outputs,

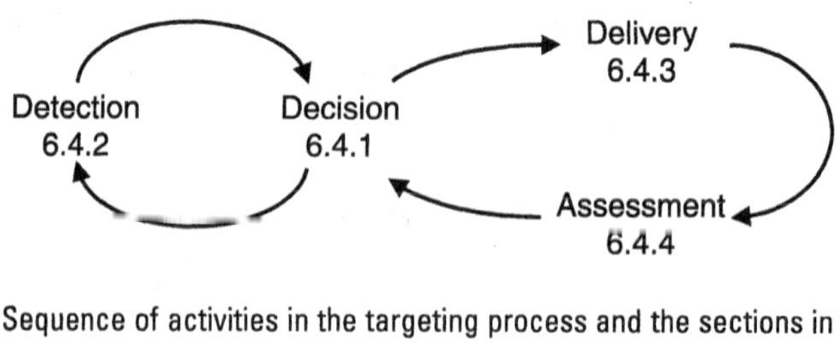

Figure 6.4 Sequence of activities in the targeting process and the sections in which they are discussed.

including target lists and taskings for the EW assets employed. While EA assets would usually be tasked to support a specific operation, ES assets will have continuous tasking to provide information on the adversary electronic order of battle. The decision process is also continuous, therefore, and cannot be simply turned on at the beginning of an operation.

The decision process can be broken up into a number of parts. These are illustrated in Figure 6.5.

The decision process runs parallel to the planning process described previously. Most of the elements of the decision process form part of the course-of-action development and course-of-action analysis processes.

6.4.1.1 Initial Plan

Planning begins with mission analysis. Initial planning proceeds to identify potential targets that are relevant to the aims of the mission. This process requires a range of intelligence about the adversary, a database of which would usually be constructed and maintained on a continuing basis. Without this database, initial planning is almost impossible. An attempt may be made to assign a value to targets in terms of their importance to possible adversary courses of action. This might involve identifying nodes in target communications networks that are critical to their operation and which, if jammed, will cause maximum disruption.

6.4.1.2 Target Development

The target development process takes information on targets from the initial plan, and coordinates targets and effects to achieve the mission. It is primarily associated with the course-of-action development, analysis, and comparison phases of the planning process. An important part of this process is the identification of high-payoff targets, where the payoff is evaluated in the context of the mission. High payoff may be due to a positive impact on

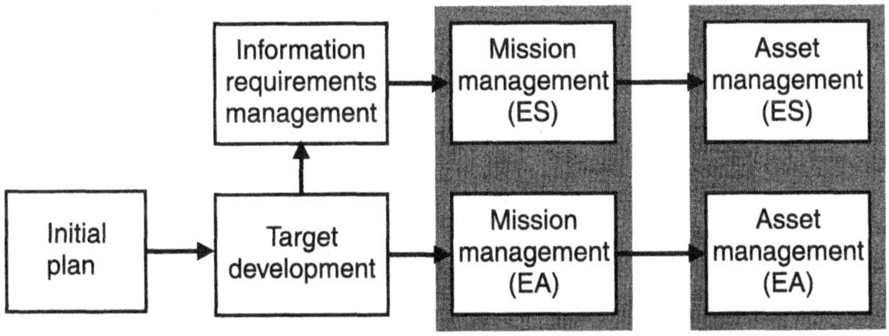

Figure 6.5 The decision process.

friendly courses of action or a negative impact on adversary courses of action. This inherently requires coordination with other parts of the operational plan.

The target development process includes the identification of high-payoff targets and the effect required for each of these targets. As with other parts of the targeting process, effects are viewed in the context of the adversary force, rather than the communications system or information system that may be the direct target of EA.

Like any other tactical action, a number of different effects can be defined for EA, including disrupt, delay, divert, limit, and destroy.

- *Disrupt.* Aims to fragment the adversary's communications system, and slow or stop the flow of information. Disruption requires relatively little coordination to implement, although deconfliction is required to ensure that the links targeted for disruption are not also subjects of an ES collection plan.
- *Delay.* Designed to delay movement and alter the time at which adversary forces arrive at a location in the battlefield. EA may be used to cause delay itself, or to deny an adversary the ability to use communications to overcome delay caused by other means. A successful application will force an adversary to use an alternative, slower means of communication, such as runners or hand signals.
- *Divert.* Aims to prevent an adversary from using critical resources. It may be applied in jamming the communications between combat units and combat support or combat service support elements, reducing the efficiency with which the adversary commander can support combat troops. This is difficult to achieve because of the level of intelligence required to pinpoint targets.
- *Limit.* Reduces the adversary commanders' options or available courses of action. In the context of EA, denial of part or all of the electromagnetic spectrum can be a form of limiting. For example, an electronic minefield of unattended jammers may be used to limit an adversary's options in the advance.
- *Destroy.* Destruction of a particular adversary combat capability is the highest level of effect that can be achieved. Destruction may be achieved through neutralization, but is unlikely to be achieved directly by either jamming or deception.

In all cases, there is a tradeoff between impact and cost; in general, the higher the impact, the higher the cost. High impacts, such as denying the electromagnetic spectrum to an adversary commander, can usually only be sustained for short periods.

Target development will also generate information requirements. For example, a radio net cannot be jammed deliberately without knowing its operating frequency and the approximate locations of its stations.

6.4.1.3 Information Requirements Management

The commander's information requirement, along with guidance on priority, is generated by the target development process. This requirement is then further refined so that it is expressed in a form that is directly collectible. A commander's information requirement may state the need to identify when an adversary attack is about to start. This may be refined to a requirement to search for the adversary's command net and monitor this net for activity indicating the lead-up to an attack. Information requirements management is primarily associated with the course-of-action development, analysis, and comparison phases of the planning process. The detailed prioritization of information requirements is also executed.

The information requirements developed provide the focus for the intelligence collection plan. ES provides one of the main means by which intelligence may be collected. Other intelligence sources may assist with details of communications networks and systems, equipment, and cryptographic codes.

Information requirements management specifies *what* to collect, as well as *when* and *where.* As well as the planning functions already discussed, information requirements management is responsible for reporting of information collected, conducting quality control, and maintaining a database of collected information. Some requests may be met from the existing database without further collection.

6.4.1.4 Mission Management

Mission management tasks EA assets (for targeting) and ES assets (for the collection of information or to support the delivery of EA). Mission management is concerned primarily with the course-of-action approval and orders-production phases of the planning process. For ES, mission management defines how the information requirements generated by the information-requirements-development process are to be collected. The input to mission management is the information requests in a form that can be directly collected. For EA, mission management translates the combat effect specified in the target development into a direct effect of an EA system.

Mission management takes the input tasks and assigns them to collection assets. In the case of ES, these assets are search, intercept, and DF systems. Mission management chooses the most suitable asset. This may involve choosing between a ground and airborne ES system. Mission management verifies the coverage of the system and chooses details such as the location from which collection will take place. These details can then be coordinated with the operation of other systems.

In tasking EA assets, the mission-management process provides a mission that specifies both the task and a commander's intent. Such a task may be to jam a specified CNR net in order to force the operators to switch from encrypted traffic to plaintext. Both the task and the intent at this level are oriented toward the immediate impact of EA on the target system.

6.4.1.5 Asset Management

The asset management process produces a chain of orders for assigning particular EA and ES assets to tasks, based on the taskings developed in mission management. The first of these is associated with the broader operational order, usually as an EW annex to the operational order. During an operation, various amendments will be made to these orders to take into account changing circumstances, primarily due to contact with the adversary. Asset management deals with the platform-specific aspects of tasking a particular asset, such as traveling times and routes for a ground-based asset.

Asset management is responsible for ensuring that continuous EA and ES coverage is provided, taking into account downtime caused by the movement or replenishment of assets. For many ground-based assets, this downtime will occur whenever the asset is moved. For airborne assets, downtime may occur during air-to-air refueling, or when the asset is required to land for refueling or maintenance.

Asset management is concerned primarily with the course-of-action approval and orders production phases of the planning process.

6.4.2 Detection

The detection function, as it applies to EW, is carried out by ES assets. It involves the analysis of the outputs of search, for example, to find frequencies being used and provide basic classification; intercept, to provide more detailed classification and traffic data; and DF, to identify the locations of transmitters.

The detection process is oriented specifically to meeting the information requirements developed in the planning process. An ongoing part of detection is generating an adversary EOB, building a database of information that can be used to satisfy future information requirements.

6.4.3 Delivery

The delivery of EA, whether in the form of deception, jamming, or neutralization, is directed by the taskings evolving from the decision process.

The use of EA always requires continuous steerage provided by ES. This steerage is analogous to the adjustment of fire for artillery that is provided by a forward observer. The ES equipment used may be integral to the EA system or may be separate, requiring separate tasking. At this level, the effects of the EA will be assessed in terms of impact on the target communications or electronic system. This will ensure, for example, that jamming is occurring with sufficient power to disrupt a particular target net.

6.4.4 Assessment

Assessment provides a higher level of evaluation than that provided in the delivery phase. Making this judgment often requires information from more than one source, or from a different source than that used to provide steerage to EA in the delivery phase. The aim of assessment is to determine the success of EA (and other fires with which it may have been combined) in terms of the effect that was specified in the target development process, which is expressed in terms of impact on an adversary combat capability. This is in contrast to the adjustment performed during the delivery phase, which for EA is expressed in terms of its effect on the system being targeted.

Assessment will often make use of intelligence gained from a number of sources. The outputs of the assessment process are battle damage assessment and a recommendation on reattack. Battle damage assessment specifies in detail the effect of the attack, not just on particular systems, but on adversary combat capabilities. Possible reattack recommendations include *reattack with the same asset, reattack with a different asset,* and *do not reattack.* Reattack with the same asset may be used to maintain the target effect (e.g., where a jammer is being used to disrupt an adversary capability). Reattack with a different asset, or with the same asset but with different operating parameters, may be used where a deception operation may not be successful in disrupting an adversary operation and a recommendation to change to jamming is made. Do not reattack might be used when an attack has achieved its aim.

6.4.5 Siting Considerations for Jamming Facilities

In addition to equipment that directly supports jamming, a jamming facility will normally include communications equipment that forms part of the EW communications system. It may be collocated with other EW (ES or EA) assets. The following criteria should be taken into account in the siting of jamming facilities:

- *Coverage.* A jamming facility's coverage of potential locations of relevant adversary receivers should be maximized with a limited number of friendly jamming assets. This coverage is determined by the area over which a sufficiently high JSR can be achieved. The size of this area will depend on the location and power output of both the jammer and transmitter being jammed. It is also desirable that receivers can be jammed individually, rather than only being able to jam a whole net. This may be achieved by careful siting of jamming facilities or the use of directional antennas on jammers.
- *Fratricide.* Jamming facilities should be sited to minimize interference to friendly communications. The use of directional antennas assists greatly here, but it does imply that jammers are located forward to reduce the impact on friendly systems.
- *Tactical situation.* Because they transmit large amounts of power, jamming facilities are often easy to locate through DF. They should be sited to minimize the vulnerability to an adversary's weapons systems. For a standoff jammer, this may mean siting out of the range of artillery (although this may significantly reduce the effective range of the jammer). For an unattended jammer, this may be achieved by placing the jammer sufficiently close to an adversary's facilities so that destruction by weapons systems cannot be used without endangering the facility being protected.
- *Communications.* Jamming facilities should be sited so that they can communicate directly with related EW command and processing and supporting ES facilities. These communications would normally be via a data-enabled VHF CNR. HF CNR may be used where the coverage of VHF CNR is insufficient.
- *Access.* The site must be capable of access using available transport, given the tactical situation. This applies both to ground-based and aircraft-based jamming facilities. Ground-based jammers must be able to tear down and depart rapidly if under attack.

- *Resupply.* The site must be able to be resupplied.
- *Defense.* A ground-based jammer is typically manned by a small crew of three to four, which has a very limited capability for self-defense. Such a jammer must therefore be sited unobtrusively or collocated with some other larger facility that can protect itself.

6.4.6 Deception Planning

In addition to the basic considerations for planning EA, the planning of electronic deception must be coordinated with the force's overall deception plan. This requires the integration of electronic deception with a range of other actions, including visual, sonic, and olfactory. The results of the adversary's electronic reconnaissance must be consistent with data obtained from other sources. For this reason, electronic deception is rarely used in isolation, and where it is, the effect is often short-lived.

In all deception operations, time is critical. Given sufficient time, the adversary can discover even the most complex electronic deception. The longer it is required to deceive an adversary, the better coordinated the electronic deception must be.

Adversary EW capabilities are a critical factor in planning electronic deception. All transmissions forming part of the electronic deception must have sufficient power to be received at an acceptable quality by the target receivers, transmitted on a frequency that the adversary is capable of intercepting and in a form that can be intercepted. There is little value, for example, in planning a deception requiring the adversary to read a particular message, and then transmitting this message using a cipher that the adversary cannot cryptanalyze. Additionally, if the adversary is expected to employ EA against dummy nets, the deception must be elaborate enough to ensure that adversary ES will be able to observe the effect that jamming would have on target nets.

6.4.6.1 Siting Considerations for Deception Facilities

In addition to equipment that directly supports deception, a deception facility will normally include communications equipment that forms part of the EW communications system and will typically be manned by specialist EW personnel. It may be collocated with other EW (ES or EA) assets. The following criteria should be taken into account in the siting of deception facilities:

- *Coverage.* A deception facility's coverage of potential locations of relevant adversary receivers should be maximized with a limited number of friendly assets. It may be desirable that receivers can be targeted individually, rather than only being able to target a whole net. This is achieved by the careful siting of deception facilities, or the use of directional antennas on deception transmitters.

- *Signature.* A deception facility may need to be sited so that its electronic signature, including DF, matches the deception plan.
- *Communications.* Deception facilities should be sited so that they can communicate directly with related EW command and processing and ES facilities. These communications would normally be via a data-enabled VHF CNR. HF CNR may be used where the coverage of VHF CNR is insufficient.
- *Access.* The site must be capable of access using available transport, given the tactical situation. This applies both to ground-based and aircraft-based intercept facilities.
- *Resupply.* The site must be able to be resupplied.

6.5 Collection Management

Collection management exists side-by-side with the targeting process, and is aimed primarily at the tasking of ES assets. The purposes of ES are tactical intelligence collection, providing steerage for EA, and cuing surveillance and target acquisition resources. ES planning supports these purposes, which are crucial to the detection and assessment functions of EA planning.

ES planning is driven by the targeting process, and forms one part of the intelligence collection management process.

6.5.1 The Collection Management Process

The collection management process is shown in Figure 6.6. The collection management process can be seen as a subset of the targeting process, incorporating those elements concerned with developing specific information requirements and tasking assets to meet these requirements.

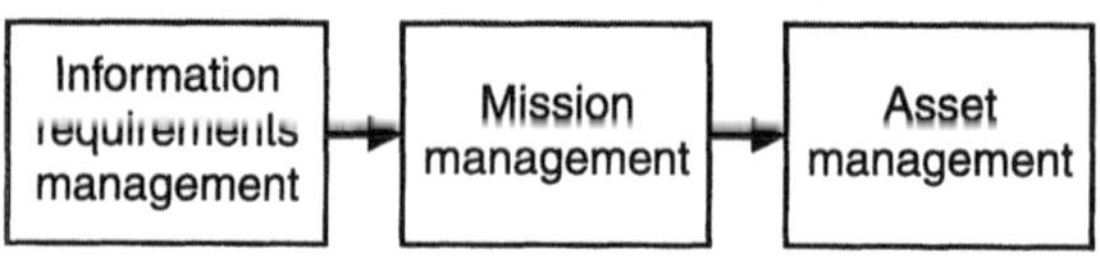

Figure 6.6 The collection management process.

6.5.2 Siting Considerations for Search Facilities

In addition to equipment that directly supports the search function, a search facility will normally include communications equipment that forms part of the EW communications system. It may be collocated with other EW (ES or EA) assets, particularly intercept facilities. The following criteria should be taken into account in the siting of search facilities:

- *Coverage.* A search facility's coverage of potential locations of relevant adversary communications assets should be maximized with a limited number of friendly search assets. At VHF and higher frequencies, the range of search receivers, fundamentally limited by terrain, will be similar to communications receivers, although a small increase in range may be achieved due to the higher sensitivity of the search receiver.
- *Communications.* Search facilities should be sited so that they can communicate directly with related EW command and processing, DF, intercept, and EA facilities as well as other search facilities. These communications would normally be via a data-enabled VHF CNR. HF CNR may be used where the coverage of VHF CNR is insufficient.
- *Access.* The site must be capable of access using available transport, given the tactical situation. This applies both to ground-based and aircraft-based search facilities.
- *Resupply.* The site must be able to be resupplied.
- *Defense.* A ground-based search facility is typically manned by a small crew, which has a very limited capability for self-defense. Such a facility must be sited unobtrusively or collocated with some other larger facility that can protect itself.

6.5.3 Siting Considerations for Intercept Facilities

In addition to equipment that directly supports the intercept function, an intercept facility will normally include communications equipment that forms part of the EW communications system. It may be collocated with other EW (ES or EA) assets. The following criteria should be taken into account in the siting of intercept facilities:

- *Coverage.* An intercept facility's coverage of potential locations of relevant adversary communications assets should be maximized with a limited number of friendly intercept assets. The prior use of search and DF may allow intercept receivers to have a reduced coverage if the intercept facility is not collocated with other ES assets.

- *Communications.* Intercept facilities should be sited so that they can communicate directly with related EW command and processing, search, DF, and EA facilities as well as other intercept facilities. These communications would normally be via a data-enabled VHF CNR. HF CNR may be used where the coverage of VHF CNR is insufficient.
- *Access.* The site must be capable of access using available transport, given the tactical situation. This applies both to ground-based and aircraft-based intercept facilities.
- *Resupply.* The site must be able to be resupplied.
- *Defense.* A ground-based intercept facility is typically manned by a small crew, which has a very limited capability for self-defense. Such a facility must be sited unobtrusively or collocated with some other larger facility that can protect itself.

6.5.4 Siting Considerations for DF Facilities

In addition to equipment that directly supports the DF function, a DF facility will normally include communications equipment that forms part of the EW communications system. It may be collocated with other EW (ES or EA) assets. The following criteria should be taken into account in the siting of DF facilities:

- *Coverage.* A DF facility's coverage of potential locations of relevant adversary communications assets should be maximized with a limited number of friendly DF assets.
- *DF location.* The DF antenna must be sited to have a clear path to the emitter and must be well forward on a feature and clear of obstruction. This often conflicts with tactical requirements. The accuracy with which the location of the emitter can be determined will depend on the ability to accurately determine the location of the DF sites. The use of GPS can dramatically improve the accuracy of location.
- *Site error.* A DF facility should be sited to minimize sources of error, such as multipath error, that depend on its surroundings.

- *Communications.* DF facilities should be sited so that they can communicate directly with related EW command and processing, search, intercept, and EA facilities as well as other DF facilities. These communications would normally be via a data-enabled VHF CNR. HF CNR may be used where the coverage of VHF CNR is insufficient.
- *Access.* The site must be capable of access using available transport, given the tactical situation. This applies both to ground-based and aircraft-based DF facilities.
- *Resupply.* The site must be able to be resupplied.
- *Angle between bearing lines.* The DF baseline should be sited so that the angle between DF bearing lines is as close as possible to 90°. Angles less than approximately 45° will lead to significant uncertainty in the transmitter location.
- *Defense.* A ground-based DF facility is typically manned by a small crew, which has a very limited capability for self-defense. Such a facility must be sited unobtrusively or collocated with some other larger facility that can protect itself.

6.5.5 Siting Considerations for Analysis Facilities

The considerations for siting an analysis facility are:

- *Communications.* Analysis facilities should be sited so that they can communicate directly with related EW command and processing facilities, as well as the search, DF, intercept, and EA facilities that are providing the data to be analyzed. This communication would normally be via a data-enabled VHF CNR. HF CNR may be used where the coverage of VHF CNR is insufficient.
- *Access.* The site must be capable of access using available transport, given the tactical situation. This applies both to ground-based and aircraft-based facilities.
- *Resupply.* The site must be able to be resupplied.
- *Defense.* A ground-based analysis facility is typically manned by a small crew, which has a very limited capability for self-defense. Such a facility must be sited unobtrusively or collocated with some other larger facility that can protect itself.

Endnotes

[1] U.S. doctrine for EW planning is contained in:
U.S. Army Field Manual FM 6-20-10, "Tactics, Techniques, and Procedures for the Targeting Process," May 1996.
U.S. Army Field Manual FM 34-2, "Collection Management and Synchronization Planning," March 1994.
U.S. Army Field Manual FM 34-45, "Tactics, Techniques, and Procedures for Electronic Attack," June 2000.

[2] U.S. Army Field Manual FM 6-20-10, "Tactics, Techniques, and Procedures for the Targeting Process," Appendix B, May 1996.

[3] U.S. Army Field Manual FM 34-130, "Intelligence Preparation of the Battlefield," July 1994.

7

Radio Frequency Directed Energy Weapons

7.1 Introduction

The most commonly used forms of EA have traditionally been jamming and deception, largely because it has been difficult to generate sufficient power levels for neutralization. Radio frequency directed energy weapons (RF DEW) have the potential to generate sufficiently high power levels to provide effective neutralization of a broad range of military electronic equipment. In principle, any equipment that employs modern electronic components is at risk from RF DEW attack. The impact of such an attack could include both disruption of the operation of equipment or destruction of electronic circuits, causing armored vehicles and ships to operate erratically or become completely inoperative, and aircraft to fall out of the sky.

In one sense, the threat from RF DEW is not new; protection against the nuclear electromagnetic pulse (N-EMP) resulting from a nuclear explosion has been recognized as important for many years. What is more recent, however, is the potential to build nonnuclear RF DEW. The principal advantages of RF DEW are that the destructive energy of the weapon is delivered almost instantaneously and many targets can be engaged at the same time. The main limitation of RF DEW is that, unlike lasers, it is not possible to produce a narrow, high-powered, focused RF beam. It is therefore very difficult to provide sufficient power on a target to achieve damage at useful military ranges, and the weapon is difficult to employ without causing collateral damage to the weapon platform or friendly forces.

The idea of using electromagnetic energy as a weapon goes back to the early part of the twentieth century. In 1924, the British government offered a prize of £1,000 for a "death ray" that could kill a sheep at 100 yards [1]. As far as is known, the prize was never claimed. Nikola Tesla claimed in 1933 to have invented a "death ray," the range of which was reported to be 200 miles [2].

There are many reports of the existence of RF DEW and of ongoing research into development and countermeasures. In 1987, the Soviet Union was reported to have conducted extensive research into RF weapons, and to be likely to test a ground-based radio-frequency weapon capable of damaging satellites in the 1990s [3]. Russian technology in the explosive generation of very-high energy RF pulses, with energies as high as 100 MJ, led U.S. technology by a factor of 10 in the early 1990s [4]. *The Washington Post* reported in 1996 that the Soviets used such weapons to kill goats at short ranges. In 1997, it was assessed that Russia and Ukraine had significant RF weapons capabilities [5].

The United States has also invested significant resources into research into RF weapons, possibly as early as the 1960s. While much of the development of explosive RF generators occurred in the Soviet Union, it is believed that the first report of this approach was that by Fowler [6]. U.S. efforts in this area emphasize countermeasures as well as weapon development. Locations at which this work takes place include the Philips Laboratory at Kirtland Air Force Base and Los Alamos National Laboratory [7], both in New Mexico. While the Soviet Union may have had led the United States in this area previously, by 1993 it was being reported that "there is little doubt that the U.S., abetted by recently bought Russian technology, has taken a considerable world lead in this field" [7].

In 1996, an Air Force Scientific Advisory Board report on future weapons included a section on an RF gunship. Additionally, the Air Force's Armstrong Laboratories at Brooks Air Force Base is heavily engaged in such research. Budget documents show that the laboratory intended to spend more than $100 million over the six years to 2003 to exploit "less-than-lethal biological effects of electromagnetic radiation for Air Force security, peacekeeping, and war-fighting operations" [8].

Major areas identified for technological development in the late 1990s [9] were compact, high-power microwave (HPM) sources weighing approximately 500 lbs and with a volume of 1.5 cu ft (excluding antenna and pulse power generator); compact, high-power, high-gain, ultra-wideband antennas approximately 18 inches in diameter with 15 to 20 dB gain; compact, efficient, high-power, pulse generator, weighing approximately 500 lbs with volume less than 10 cu ft and peak power greater than 50 GW; development of explosively driven, pulsed-power sources; systems integration into existing military platforms, including aircraft, land vehicles and ships; and radiation hardening of existing assets.

Funding for development of RF DEW in the United States includes $19.8 million in 1996 for high power microwave technology, with a further $63 million planned over the following seven years [10]; $71.5 million between 1997 and 2003 for "High-Power Microwave C2W/IW Technology" to "disrupt, degrade, and destroy electronics in communication and information systems to support command and control/information warfare (C2W/IW) and suppression of enemy air defense (SEAD) missions" [11]; $18.5 million in 1996 with $88.5 million projected between 1997 and 2001 for HPM/LASER aircraft self protect missile countermeasures [12]; and a $6.6 million contract let to Hughes Missile Systems for demonstration of "the high power microwave suppression of enemy air defense (SEAD) technology" [13].

There have been a number of reports of U.S. RF DEW capabilities. The U.S. Air Force is reported to have modified cruise missiles to take what is surmised to be an RF DEW [14]. In 1999, the Los Alamos National Laboratory was reported to have developed a high-power microwave weapon capable of being carried in a laser-guided bomb or cruise missile [15]. This device was reported to have a power output of "tens of terrawatts," and a range of up to 50m. A further indicator of the importance placed on RF DEW by the United States is the explicit inclusion in 1982 of monitoring of nonnuclear electromagnetic pulse (EMP) sources in the functions of the SIGINT committee [16].

Interest in RF weapons has also been evident in The Netherlands, where the Royal Netherlands Navy has examined the potential impact of these weapons on future ship design, from both offensive and defensive perspectives [17]. The United Kingdom and France are reported to have significant RF source development efforts [5], with the United Kingdom also studying a variety of RF weapons [7]. Lesser efforts have been noted in Germany, Switzerland, China, Japan, Sweden, Israel, and Australia [5].

The aim of this chapter is to examine the potential impact of RF DEW, and in particular nonnuclear RF DEW, against land forces. It begins with a brief discussion of the properties of RF DEW, followed by an examination of possible platform protection mechanisms and their likely effectiveness. The power levels required to achieve particular target effects are then analyzed, and the maximum ranges at which these effects can be achieved. Delivery mechanisms are then discussed, leading to an analysis of the threat posed by these weapons to land forces, relative to that posed by conventional weapons. Finally, an analysis is presented of the planning considerations for defense against use of RF DEW by an adversary.

7.2 Characteristics of RF DEW

RF DEW operate by projecting electromagnetic energy from a transmitter onto a target at a sufficient power that electronic systems are damaged or their operation disturbed. Damage is caused to a wide range of electronic equipment because components, wires and apertures act as antennas to couple the weapon's energy into the target. RF DEW, therefore, have the potential to affect almost all electronic equipment, not just communications receivers that might happen to be tuned to the RF DEW transmission.

7.2.1 Continuous Wave and Pulsed RF DEW

While there is no agreed definition, RF DEW are herein considered to be high-powered transmitters (transmitting up to 10 GW) that can produce frequencies within a very wide portion of the electromagnetic spectrum (up to 100 GHz). The transmitter may produce continuous power, known as *continuous wave* (CW), or it may produce one or more short pulses. The major uses of CW are jamming, and to target personnel. These uses are not considered further in this chapter.

Pulsed RF DEW could be used against a large variety of electronic equipment, where the damage caused might be transient (for example, introducing errors into a computer) or permanent (such as the destruction of electrical conductors within a circuit, preventing it from functioning correctly). These weapons include electromagnetic pulse (EMP) and high-power microwave (HPM). EMP weapons aim to produce a broad band of frequencies, usually covering frequencies at 10 MHz and below. EMP weapons are reported to have no effect on personnel [18]. HMP weapons tend to produce narrowband pulses in the microwave band, with pulse lengths up to 1 ms.

7.2.2 Explosive and Nonexplosive RF DEW

Pulsed RF DEW can be divided into two categories: *nonexplosive* and *explosive weapons.*

Nonexplosive RF DEW produce one or more pulses of RF energy. Typical nonexplosive RF DEW would generate long chains of pulses with some predefined repetition time. Such weapons require a large platform due to the weight and space requirements of the power generator. In the discussion below, we will use the example of a single RF DEW with a peak power of 10 GW, a pulse length of 1 μs and a repetition time of 1s.

As shown in Figure 7.1, the nonexplosive RF DEW consists of the following components: a continuous power generator, a pulse generator, an RF converter, and an antenna. Some form of cooling is also usually required.

Continuous Power Generator A source of continuous power is required for operation of the RF DEW. This may take the form of a battery bank, a diesel generator or be derived from a vehicle's internal power. The average power supplied must be greater than the average power required by the pulse generator to allow for losses in the system. For our example 10-GW system, an average power of 10 kW is required. In practice, it is unlikely that the system will be more than 10% efficient, so 100 kW are probably required.

Pulse Generator A pulse generator stores energy produced by the continuous power generator and outputs this energy in short pulses. For our example system, a volume of approximately 0.1 m^3 may be required with weight up to 100 kg. In order to minimize losses, the pulse generator should be located as close to the RF converter as possible.

RF Converter The RF converter takes the output from the pulse generator and produces the signal to be transmitted via the antenna. In a nonexplosive RF DEW, such as a high power microwave device, the RF converter could be a magnetron that modulates the pulse to shift its frequency into the microwave band. A practical weapon system might have a volume of approximately 2×10^{-3} m^3 and a weight of approximately 40 kg.

Antenna The antenna efficiently couples energy from the RF converter to the surrounding atmosphere. Its size will depend on the type of antenna.

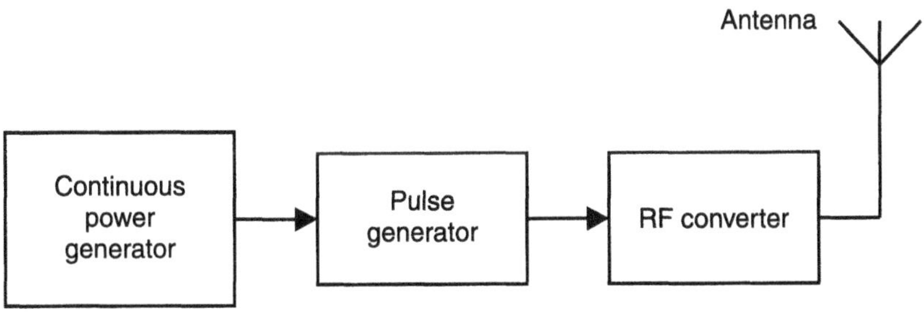

Figure 7.1 Block diagram of nonexplosive RF DEW.

Suitable antennas may include a 1m-diameter parabolic dish weighing up to 10 kg. The antenna must be located in a suitable position to illuminate targets. Likely locations are the nose of a missile, the top of a mast, and the underbelly of an aircraft.

Cooling A one-shot RF DEW would not usually require a significant cooling system. A continuous-pulse system, however, may require a large cooling system. Much of this cooling will be required in the pulse generator and RF converter stages. Based on the previous example, and assuming an overall efficiency of 10%, the cooling system would have to dissipate 90 kW.

Explosive RF DEW produce a single pulse of energy, derived from the detonation of an explosive charge. Since the energy for these weapons comes from a small explosive source, they are generally much smaller than their nonexplosive counterparts and can be deployed by missile. Explosive RF DEW are also most suitable for missile applications because of the damage that is likely to be caused to the weapon platform on detonation. Explosive RF DEW may be generated by either a nuclear or conventional explosion. This chapter concentrates on the nonnuclear case.

7.3 Target Effect Mechanisms

Because of the difficulty of delivering large amounts of power over long distances, the major damage mechanism of RF DEW is not destruction of a target, but rather the damage achieved by penetration of the system by the RF energy to reach some key, sensitive components in the target. The destruction of these key components then corresponds to destruction of the equipment.

Target-effect mechanisms of RF DEW are generally divided into two broad classes: front-door damage through deliberate antennas used by the target for communications or RADAR, and back-door damage, where radiation is coupled through nondeliberate antennas such as lines and component legs, as well as through apertures and gaps in seals.

7.3.1 Front-Door Damage

Front-door damage occurs in systems, such as communications receivers and radar, which have high-gain antennas that are designed to collect electromagnetic radiation at a selected frequency. An RF DEW can couple its emissions at this frequency most effectively into the target when the antenna collects the RF DEW signal and focuses it into the target's own receiving subsystems. Therefore, front-door damage has much in common with jamming, since its performance relies on power being coupled into a receiver through a communications antenna.

7.3.2 Back-Door Damage

Back-door damage occurs when the RF DEW signal is coupled to components and subsystems. The signal can enter the target through small apertures or seams in subsystem enclosures and interconnecting cables. The gaps act like waveguides to funnel the RF DEW signal into the electronics boxes, which then act as resonant cavities for the signals. In addition, interconnecting cables act as simple antennas. Once the electromagnetic radiation is inside the equipment enclosures, various electronic components act as simple monopole antennas that will couple to signals whose wavelength is roughly equal to the length of the component. Consequently, resistors, wires, printed circuit-board interconnects, component legs or connections all act as good collectors of RF DEW signals, which are then fed through the target-system electronics to disrupt the more sensitive elements. Clearly this type of attack requires higher power densities to achieve the same level of damage possible via the front-door approach since it does not have the benefit of the high-gain antenna to collect and focus the signal.

The components that are the most vulnerable to RF DEW are the sensitive semiconductors (such as receiving diodes and logic chips) contained in modern equipment such as radar and communication sets and electronics-based equipment (particularly those that are computer-based). Higher levels of signal can affect the electronics in ignition systems, and at very high levels it is even possible to detonate missile warheads, bombs, and possibly artillery shells.

The main threat is from an RF DEW that can deliver its energy in about 1 μs or less since semiconductor junctions need this amount of time to dissipate the heat. The higher the state of technology, the more vulnerable it is. Large-scale use of solid-state components and the increasing sophistication of component circuitry have increased vulnerability of systems. This is particularly so when large-scale integrated circuits are used. It is important to recognize that this trend cannot be reversed without dramatically reducing the functionality of modern platforms. The solution lies, therefore, not in returning to the simpler circuit designs of the past, but rather in incorporating hardening into circuit and equipment design.

7.3.3 Damage Mechanisms Employed by RF DEW

Once large amounts of RF energy have entered an electronic system, the system's performance can be degraded in a number of different ways, including component failure due to mechanical defects, equipment upset leading to a temporary failure, insulation breakdown, leading to either temporary failure or destruction, failure of semiconductor junctions, and burnout of metal interconnections.

In all cases, it can be expected that the greatest impact will occur in equipment, such as commercial-off-the-shelf equipment, that has not been designed to operate in a hostile electromagnetic environment.

The susceptibility of systems and subsystems depends on the constituent devices. For example, a receiver that uses valves has a moderate susceptibility, whereas a receiver built from semiconductor devices has a high vulnerability [19]. The vulnerability of a variety of different components is shown in Figure 7.2. Typical values relevant to the operation of computer and communications equipment include 80 mW to disrupt the operation of an unshielded computer, and 1W to destroy an unprotected transistor in the input stage of a radio receiver.

7.3.4 Propagation Effects

Once the RF DEW energy has been launched from the antenna, it must propagate to the target. The amount of energy incident on a target is limited by three factors: radio line-of-sight, atmospheric dielectric breakdown, and atmospheric absorption.

Radio Line-of-Sight When a radio wave travels through the atmosphere, it is affected by the refractive index of the atmosphere, which tends to bend the wave towards the Earth. Lower frequencies are more affected than higher frequencies, and VHF and low UHF frequencies tend to follow the curvature of Earth and can therefore travel further than the optical horizon. At higher frequencies (> 1 GHz), the radio waves are unaffected by the refractive index profile and tend to travel in a straight line. In that case, the range of transmission will be limited by the horizon due to the curvature of the Earth. This range is called the optical horizon, which would be approximately 10 km over a flat surface for an RF DEW mounted 10m above the ground, aimed at a target placed on the ground.

Atmospheric Dielectric Breakdown Another major limiting factor to the amount of power that can be transmitted through the atmosphere is atmo-

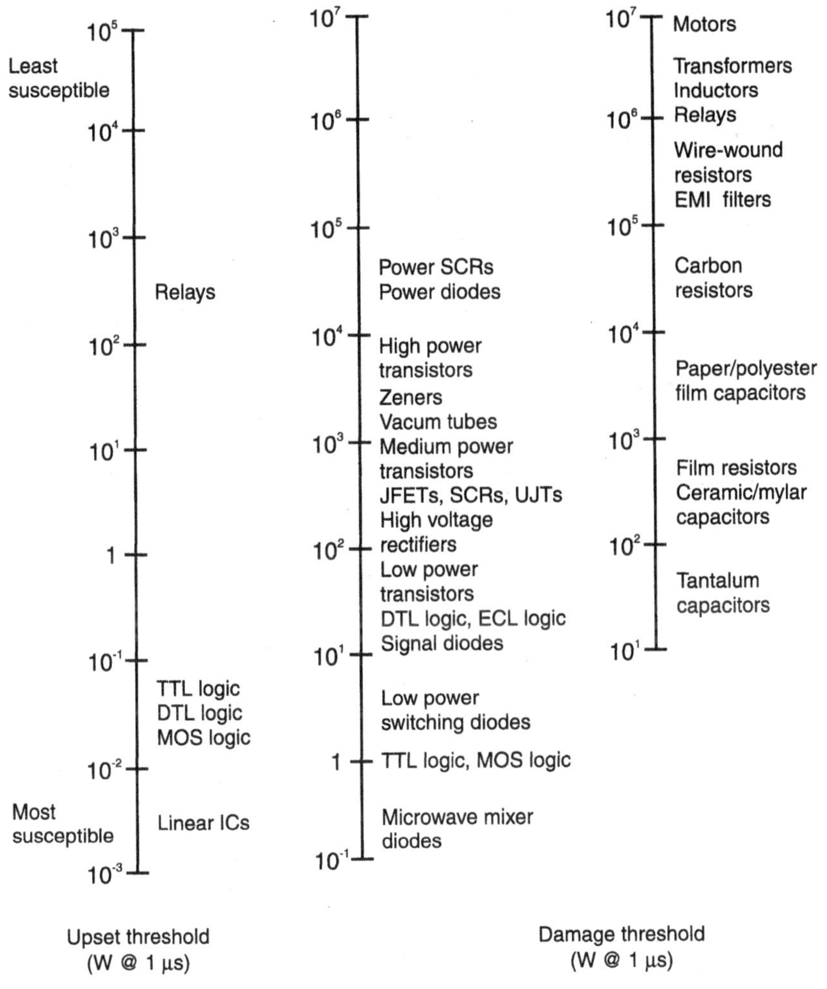

Figure 7.2 Powers required to damage various electronic devices [19].

spheric dielectric breakdown, which occurs when the high field strength of the RF DEW electromagnetic signal leads to atmospheric ionization. The field strength at which breakdown occurs depends on signal frequency and pulse length as well as on the air pressure [20]. This limit on the power output (approximately 10 GW) provides a practical limit on the range of a nonnuclear RF weapon, and is a major reason why RF DEW cannot be used to cause structural damage at any useful range.

Atmospheric Absorption Atmospheric constituents at particular frequencies absorb significant amounts of RF energy. For example, water vapor absorbs strongly at 22 GHz and 185 GHz, while oxygen absorbs at 60 GHz and 118 GHz. This has the effect that, for weapon ranges between 1 and 100 km, attenuation within several GHz of the water vapor and oxygen absorption lines will be too high. Precipitation is also a major atmospheric absorber. Absorption varies with frequency since scatter depends on the ratio of droplet size to electromagnetic wavelength. For example, over a 10-km range, 3-GHz radiation is attenuated by 0.01 dB in moderate rain, while 30-GHz radiation is attenuated by 10 dB in the same conditions [20].

7.4 Platform Protection

Having considered the potential impact of RF DEW, it is important to consider the methods available to counter them: in other words, EP. Because it is very difficult to simulate high-power pulses, the development of protection systems has not been an exact science. However, there are some agreed methods that can be employed to protect platforms, systems, and equipment. Two main techniques are avoidance of illumination by the weapon, and hardening of platforms/systems/equipment against the effects of the weapon.

7.4.1 Avoidance of Illumination

Avoidance of illumination involves ensuring that any threat carrying a single RF DEW cannot get close enough to the platform to be effective. In this respect, RF DEW are no different to other weapons systems where avoidance of detection and illumination by the system is the first level of defense. In addition, platforms may use terrain shielding as a form of EP. It is therefore a matter of tactics.

7.4.2 Hardening Techniques

Hardening involves the development of an electromagnetic barrier to prevent harmful transients from reaching sensitive equipment. There are a number of hardening steps that can be taken, but, as with any countermeasure, there is a price to pay in terms of increased size, weight, and cost. The techniques that offer the greatest value to counter RF DEW are the use of filters to reduce out-of-band coupling, the use of over-voltage or over-current protection devices to arrest pulses, connection between subsystems using fiber-optics, the replacement of sensitive components with hardened functional equivalents, incorporation of redundant circuits, the development of fault-tolerant devices, and shielding and grounding of cables, subsystems, and full systems.

MIL-STD-188-125 [21] gives general requirements for the integration of EMP hardening with other electromagnetic interference (EMI)/electromagnetic compatibility (EMC), lightning protection, and TEMPEST design requirements [22].

MIL-STD-464 [23] is utilized as the standard to establish the electromagnetic environmental effects interface requirements and verification criteria for all land, air, and sea systems, including associated ordnance. Among other things, the standard describes the maximum levels of electrical field allowable in the environments into which systems are to be deployed. These levels are used in this chapter as the threshold that would have to be exceeded by the RF DEW to cause any ill effects or damage to target systems.

Platform hardening can be approached at one or more of three layers or levels: the platform level, the system level, and the equipment interface/circuit level [24].

Platform-Level Hardening (Layer 1) Layer 1 hardening involves shielding external electrical conductors, installing protective devices at external electrical penetration points, and controlling cable routing to minimize the possibility of energy being transferred to interior spaces. EMP hardening of approximately 40 dB was achieved in hardening Aegis class cruisers [24]. It is expected that this level could also be achieved for the wider range of frequencies achieved with RF DEW. For land forces, this type of hardening is often difficult to achieve. Armored vehicles, for example, are not traditionally designed with the necessary shielding. In addition, much equipment, such as Combat Net Radio, must be operated from platforms, such as soldiers' backs, that are not capable of being hardened.

System-Level Hardening (Layer 2) This layer addresses shielding interrack connecting cables, grounding shields at connector backshells, and shielding/grounding of equipment racks to reduce the possibility of DEW energy being transferred to equipment interface circuits. Terminal protection devices have been reported to provide an additional 15-dB protection between HF/VHF antenna feeds and receivers [24]. For land forces, this type of hardening is feasible, provided that it is taken into account during system design.

Equipment Interface/Circuit-Level Hardening (Layer 3) Hardening at this level involves using balanced, high common-mode rejection interface circuits, replacing interrack signal cables with fiber optic links, incorporating protective devices on interface lines, circuit boards, or chips, and selecting harder circuit components to assure that the equipment can withstand any residual RF DEW energy. Protective devices need to be individually designed for each system or circuit to be protected. An example of integrated overvoltage protection for an HF receiver is shown in Figure 7.3, incorporating spark-gap devices for coarse protection and a varistor and diodes for fine protection.

7.4.3 Issues Associated with the Protection Provision

Protection Planning The main difficulty in providing protection against the effects of electromagnetic radiation, whether generated by RF DEW or natural sources such as lightning, is not the technology or the expenditure. Rather, it is in the equipment procurement process: sufficient planning must occur to ensure that adequate provision is made in the early planning phases for the appropriate protection of platforms, systems, and equipment. MIL-STD-464 provides a basis for determining the level of protection required at each layer.

Cost Providing that consideration is given early enough in the acquisition cycle, the cost of integrated protection (such as that espoused by MIL-STD-464) to encompass all sources of electromagnetic radiating, can cost as little as 2 to 10% [25], most of which is spent in performance testing and life-cycle maintenance of robustness against electromagnetic radiation [18]. If

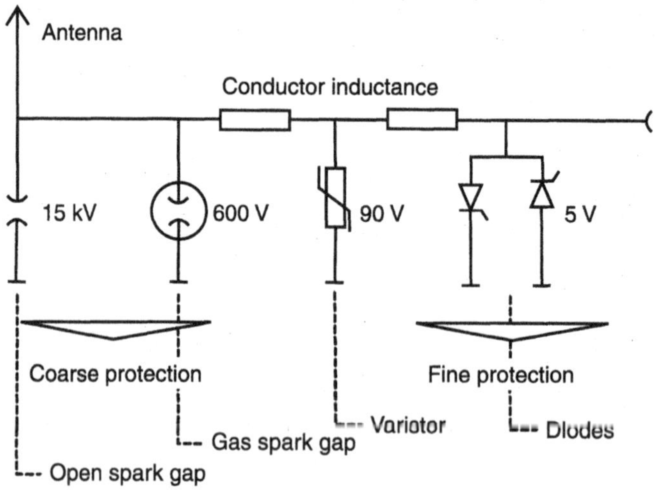

Figure 7.3 Integrated overvoltage protection for an HF receiver [25].

included in the procurement phase, the costs of integrated protection are small enough to be considered negligible, particularly in light of the platform, system, and equipment vulnerability that results if such protection is not included. Costs of hardening equipment later in its life (where technically feasible) have generally been approximately 20% of total system cost [18].

Tactics and Doctrine In a manner similar to NBC protection, it is likely that platform protection measures will result in some constraints on the operation of particular platforms. For example, armored vehicles and shelters may be designed/modified to increase sealing to reduce the ingress of RF energy. These measures will only be effective if hatches and doors remain shut. The capability of vehicles such as armored reconnaissance vehicles may be restricted if they have to operate closed-down. The impact of RF DEW on tactics and doctrine is discussed in detail in Section 7.5. Training must also take place in a realistic environment to ensure successful operations when the use of RF DEW is likely.

Maintenance The maintenance of the desired level of hardening throughout the life cycle of the platform requires a number of additional supporting activities. In particular, operators and technicians must receive additional training in the maintenance of systems, subsystems, and devices that have been modified for protection against RF DEW. The protection of a platform relies on the interaction of a range of protection levels and activities. Users and repairers must be aware of the types of protection so that the appropriate levels are not inadvertently reduced during maintenance.

Disaster Recovery Plans In addition to technical solutions for protection against the effect of RF DEW, disaster recovery plans must be developed to allow platforms to recover as best as possible, should key systems be destroyed or degraded. The plans should include reversion to manual operating procedures (where possible), incorporation of redundant equipment, and so on. While these plans should already exist in the absence of a RF DEW threat, they would need to be modified to take such a threat into account. Disaster recovery plans should also be exercised in training to ensure that all personnel are aware of their roles in regaining as much operational ability as possible following an RF DEW attack.

7.5 Analysis: Offensive RF DEW

In this section, the issues associated with the offensive use of RF DEW (i.e., EA) are analyzed; the defense against such attacks is examined in Section 7.6.

7.5.1 Delivery Factors

When considering the range of possible active applications, the following factors are taken into account:

- *Physical considerations.* The size and weight of the weapon must be considered for each type of platform. The type of weapon also needs consideration. For example, an explosively generated RF DEW is not suitable for mounting on a high-value vehicle or aircraft because of the damage that would be caused to the platform itself. Similarly, a nonexplosively generated weapon is likely to be too large to be mounted in a missile or a bomb.
- *Power.* The maximum achievable power will be limited by the characteristics of the generator, its antenna system, and losses such as ionization of air if sufficient power is generated to cause an atmospheric breakdown. The breakdown of air surrounding the antenna provides a practical upper limit to the output power of an RF DEW [26]. Therefore, power levels up to a maximum of approximately 10 GW can be obtained from explosive or nonexplosive generators.
- *Range.* The range of effect of RF DEW will be estimated in Section 7.5.2. In practice, this is unlikely to be more than 10 km and will often be much less. In the land environment, the range will most likely be reduced further by terrain effects.
- *Suicide.* Unless the weapon platform is to be sacrificed, it must be sufficiently protected to prevent self-inflicted injury as a result of firing the RF DEW. Since the platform is much closer to the transmit antenna than the intended target, it follows that the platform is potentially in a much larger electromagnetic field than the target. If the DEW antenna is mounted at a height L above a ground vehicle, then the power levels at the vehicle are R/L greater than the intensities at the target, where R is the distance to the target. Therefore, the DEW platform must be protected to the levels required by MIL-STD-464 plus $20 \log(R/L)$ dB if it is to survive the activation of the weapon. These levels can be lowered due to the reduction in power as a result of the radiation pattern of the antenna—the platform will not be in the main beam of its own weapon, but will be in one of the side lobes.
- *Fratricide.* Also significant is the potential fratricide that will result from friendly platforms within the beamwidth of the DEW antennas. There is little that can be done about this technically. The best

controls come from the implementation of standard operating procedures for the use of the weapons. Suicide and fratricide provide significant limitations on the employment of RF DEW on land, sea, and air platforms.

- *Target effect.* In all cases it is assumed that the required target effect is to exceed the values required by MIL-STD-464 on the assumption that this will result in damage at the target. If the target is less protected, damage can obviously occur at longer ranges.

7.5.2 Range of RF DEW

The electromagnetic radiation used by RF DEW to deliver energy to a target is subject to losses during propagation similar to other types of electromagnetic radiation, including those used for communications and radar. The range of RF DEW depends on a number of factors, including the altitude at which the energy is released, the altitude of the target, and the terrain between the weapon and target.

We assume that the RF DEW generates 10-GW peak power, that there exists approximately 10 dB of additional loss in the receive system, and that there are no additional losses due to ionization of air surrounding the RF DEW [27]. Given the power required to cause a particular target impact, it is then possible to determine the maximum path loss that will still provide sufficient power at the target.

The operation of a land-based electronic system shielded in accordance with MIL-STD-464 should not be subject to interference at received power levels less than 3kW at low frequencies (< 10 MHz) and 10W at higher frequencies [28]. A loss in propagation of more than 35 dB at low frequencies and 60 dB at higher frequencies, therefore, will result in insufficient energy reaching the target to disrupt its operation. Assuming no impact from terrain (i.e., all loss is due to free-space loss), this is equivalent to a distance of less than 100m between the weapon and target at low frequencies and 1 km at higher frequencies.

For the destruction of an unprotected transistor attached to a receive antenna, a power of approximately 1W is required, assuming an operating frequency of approximately 1 GHz. This power will be achieved with a path loss of 80 dB. In the absence of terrain, this will provide a maximum range of 8 km.

For interference with the operation of an unshielded computer, a received power of approximately 80 mW is required, assuming an operating frequency of approximately 1 GHz and giving a maximum path loss of

90 dB, which is equivalent to a maximum range of approximately 25 km in the absence of additional loss due to terrain. Table 7.1 summarizes these likely maximum ranges for RF DEW, based on free-space loss.

In the use of RF DEW against land forces, terrain will also have a significant impact, reducing the power delivered to a target due to terrain screening, diffraction, and reflection loss. These additional losses will be minimized by activating the RF DEW as high as possible from the surrounding terrain. As long as there is no significant ionization of air surrounding the RF DEW, these calculations will be similar to those used for communications planning.

As a general rule of thumb, at most RF DEW frequencies, line-of-sight is required between the weapon and the target. This requirement for line-of-sight reduces the effective range of RF DEW to the optical horizon. The distance between the RF DEW and the optical horizon will depend on the heights of the weapon and the target, as well as the terrain in between. For example, across perfectly flat ground, the maximum line-of-sight distance between a weapon mounted at 4m above the ground and a target on the ground is 7 km. Any terrain between the weapon and the target would block out RF energy and protect the target.

In a practical use of RF DEW, it would be necessary to conduct experiments to determine the amount of loss that should be allowed for in the receiving system. Significant advantage may also be obtained by the choice of operating frequencies to take advantage of known sensitivity of particular targets.

7.5.3 Delivery Platforms

RF-DEW might be delivered from a variety of platforms, including surface vehicles (land or sea), aircraft, missiles, foot-mounted forces (particularly special forces), and terrorists. In this section, we examine the issues associated

Table 7.1
Likely Maximum Ranges for RF DEW Based on Free-Space Loss

Target	Maximum Range
MIL-STD-464 protected, < 10 MHz	100m
MIL-STD-464 protected, > 10 MHz	1 km
Destruction of unprotected transistor	8 km
Interference with unprotected computer	25 km

with the delivery by each of these means. The related issues of protection against each means of delivery are discussed in Section 7.6.

The details of the size and weight of RF DEW generators are not available in the public domain. The minimum weight of nonexplosive RF-DEW capable of generating long pulse trains is likely to be in the range from 200 to 300 kg, although generators capable of only a single pulse could be built smaller.

Kopp [29] surmises that explosive generators would be small enough to fit into a missile or a bomb, and provides some details on how such weapons might be deployed from existing platforms.

Ground Vehicle A ground vehicle could be used to carry either explosive or nonexplosive RF DEW. In the case of the explosive generator, the vehicle itself may not survive activation of the weapon. In the case of the nonexplosive weapon, the vehicle may require special hardening to enable it to continue operating after the RF DEW has been activated. Given the impact of terrain on RF propagation, it will most likely be necessary to provide a means for increasing the elevation of the RF DEW, either by use of high ground or by means of a mast. In the case of a single-use RF DEW, a mechanism similar to that used in a jumping-jack mine might be suitable for small weapons.

Artillery and Tank Guns Given that the RF DEW projectile is likely to weigh 200 kg or more, it is unlikely that conventional artillery would be a suitable launch platform. Large naval guns may be capable of performing this function if land forces are sufficiently close to the coast. Regardless of size and weight constraints, it is unlikely that the RF DEW weapon could withstand the forces involved in firing such projectiles over long distances. In this situation, it makes sense to think of the operating range of RF DEW as similar to the concentration of artillery; it provides a method for employing the weapon for area coverage, without necessarily knowing the precise location of the target.

Special Forces RF DEW may be delivered by special forces carrying the weapon to a point that is within range of the intended target. The weight of current weapons (likely to be of the order of 200 kg) makes this difficult, unless the weapon can be disassembled into a number of man-portable parts. Due to the effect of terrain, it is likely that some form of elevation will be required before reasonable effective ranges can be obtained.

Terrorists In many modern conflicts, terrorists may pose a threat to deployed military forces as well as to military bases and national infrastructure. Recent examples of such operations include NATO in Kosovo and INTERFET in East Timor. Unlike the employment of special forces, RF DEW may be more easily deployed by terrorists. Small civilian vehicles could be used to deploy the weapon in a similar manner to car bombs. However, unlike car bombs, the greater range of RF DEW poses a much larger threat to military installations such as airfields and command and control centers.

Aircraft An airborne RF DEW has great advantage over a ground-based system because it operates well above the terrain. However, because of the range limitations discussed previously, it is unlikely that an airborne RF DEW would be effective against a hardened target. Furthermore, the carriage of the required generator capacity on board a fighter aircraft is not thought to be feasible, but a purpose-built RF "gunship" could be implemented if the issues of self-damage could be overcome. Such an aircraft may have the advantage of being able to stand off from a target more than a conventional ground attack aircraft. This delivery method may be used to provide suppression of enemy air defense (SEAD) prior to launching a conventional attack to destroy the target.

Missile-Borne RF DEW A missile-borne weapon is thought to be practical where the RF energy is generated as a result of an explosion. Indeed, there is evidence that cruise missiles have been modified to enable them to carry RF DEW [15]. At the extreme, if the power generated was 10 GW, the missile might be able to explode as much as 10 km from the target (assuming line-of-sight between the weapon and the target) and still achieve a significant impact. Missiles have a similar advantage to aircraft in that they elevate the RF DEW, and are therefore less affected by terrain than ground-based systems. Realistically, it is more likely that such devices would only be effective exploding at ranges of hundreds rather than thousands of meters. Given that the likely weight of an RF DEW is around 200 kg, it is unlikely that it will be possible in the near future to use either short-range or medium-range anti-armor weapons (SRAAW or MRAAW) as a launch platform. Missile-based RF DEW are therefore likely to be launched from larger ground-based or airborne platforms.

7.5.4 Target Effects

Having considered the possible delivery platforms, we now examine the possible target effects: likely recovery times and levels of damage.

The impact on the vehicle control, sensor, and weapons systems in an armored vehicle could be either disruption or permanent damage to electronics. The recovery time for disruption is likely to be on the order of seconds. Permanent damage to electronics may constitute an electronic kill, recovery from which will require repair, possibly taking several hours.

For the disruption of C2 systems in CP or logistics installation, the recovery time is likely to be on the order of minutes to reboot computers. Some data may be lost, but in a carefully designed system the impact of this loss is likely to be small. If a large IP network is used, the recovery of routing tables for routers may take longer, reducing the systems' connectivity for this period. The impact on the operation of a CP would be reduced if an alternate means was established that was not vulnerable to RF DEW, such as magneto telephones and manual C2 systems. Permanent damage to electronic systems in a CP or logistics installation may have a similar impact to the same damage to equipment in an armored vehicle. If an alternate location is established, however, the impact of such damage on C2 may be minimal.

7.6 Analysis: Defense Against RF DEW

In this section, the characteristics of RF DEW are examined from the point of view of defense. The potential range is considered first, followed by aspects specific to the delivery platforms suggested previously.

In Section 7.4, the range of RF DEW was analyzed from the point of view of having confidence that sufficient power would reach the target to achieve the desired effect. In this section, however, the issues are examined from the point of view of being confident that the operation of systems will not be compromised as long as adversary RF DEW are not activated within a certain range of the asset being protected.

In Section 7.4, conservative estimates of range were made by assuming that 10-dB loss occurred in the receiving system. In this section, we assume that there is no loss in the receiving system, making the estimates conservative with respect to defense. The maximum ranges over which an enemy RF DEW might be effective, based on the same assumptions made in Section 7.4, are then for equipment protected in accordance with MIL-STD-464: 300m for frequencies less than 10 MHz, 3 km for higher frequencies; destruction of an unprotected transistor attached to a receive antenna: 25 km; and interference with the operation of an unshielded computer: 80 km.

These figures assume that the only source of loss is free-space loss. However, additional loss will usually occur due to terrain shielding, diffraction, and reflection. In situations in which there is not a large volume of air ionized around the RF DEW, these additional losses can be taken into account by methods similar to those employed for communications planning. Further confidence in the safety of systems could be obtained by assuming that the resilience of equipment was degraded (e.g., by 10 dB), compared to that specified in MIL-STD-464.

These maximum ranges suggest that systems must be protected to the standards specified in MIL-STD-464. Furthermore, they suggest that unless terrain shielding is used in an effective manner, headquarters as high as brigade may be within range of ground-based RF DEW operating from within enemy-controlled areas. Indeed, hindering illumination by enemy RF DEW may be considered one of the advantages of a reverse-slope defense.

Another issue worthy of consideration is the identification of the location of enemy RF DEW. This problem is related to the function performed by artillery locating units and also to DF in EW. The process of locating enemy RF DEW may be significantly easier for very high power weapons that cause significant ionization of the air.

The basis for defense against RF DEW is the establishment of an exclusion zone around assets to be protected. The key issue in developing defensive strategies against the various RF DEW delivery platforms is to identify how the use of the RF DEW changes the capability of the platform compared to its more conventional weapons systems. In this section, the defense of vehicles, aircraft, safety and arming systems, personnel, support equipment command posts, and logistics facilities will be examined.

7.6.1 Land Vehicles and Shelters

Vehicles contain a number of systems that may be neutralized by RF DEW, including engine management and control systems, communications equipment, sensors, information systems, and ordnance.

Armored vehicles and communications shelters can potentially provide platform-level protection for electronic equipment. Requirements will include complete shielding of the vehicle or shelter, with special conductive seals around doors and hatches, maximum use of optic fiber for communications within and between vehicles, and provision of suitable hardening for necessary electrical interfaces, including antennas and power cables. It is important to note that this protection is only effective when armored vehicles are operating closed-down and shelters have doors shut.

Other military vehicles are less able to provide platform-level protection. Soft-skinned equipment such as logistics vehicles, engineering plant, and communications vehicles are all vulnerable and could potentially be disabled by RF DEW. Communications and other electronic equipment operated in these vehicles may require higher levels of integral hardening than similar equipment operated from hardened armored vehicles and communications shelters.

The total exclusion of potential delivery platforms from an area of a 3-km radius around a vehicle to be protected is unlikely to be feasible. This is particularly so for vehicles operating at checkpoints for maintaining an exclusion zone around command posts and logistics installations.

The tactics of armored vehicles, which attempt to make use of terrain to minimize visibility to the enemy, may already provide some protection. This will be aided by the fact that the range of modern MRAAW is similar to that of RF DEW and that line-of-sight is usually required for launching these weapons. The Javelin weapon system, for example, has a range of approximately 2 km. Other types of operation that will tend to increase protection include operation submerged in water.

Existing procedures for NBC defense, such as closed-down operation, will also be of importance in providing a measure of protection against an RF DEW threat.

7.6.2 Aircraft

Aircraft are very vulnerable to RF DEW: If attacked while airborne, platforms could simply fall out of the sky; if attacked on the ground, aircraft could be rendered inoperable for a considerable period. Detailed analysis is required to determine the methods for route planning and protection around airfields to minimize unnecessary exposure to this threat.

A modern aircraft can be highly sensitive to EMP, which can be coupled with antenna-like protuberances. Currents induced on the outer skin of the aircraft or on a deployed trailing-wire antenna can flow into the aircraft interior through apertures such as cockpit windows or imperfect seams at skin-panel joints, or along antenna mounts or other direct penetrations. A modern aircraft carries sensitive electronics and is partly built out of advanced materials, such as carbon-fiber plastics, which have a low electromagnetic shielding effect. The use of these materials also complicates the problem of EMI, since many parts of the aircraft contain no conducting and shielding metal structures. On those external surfaces that do conduct, induced currents can reach 1,000A, and currents induced on a deployed trailing wire can reach several thousand amperes. These induced currents are much greater than aircraft interior wiring currents, which typically range from less than 1A to 10A or more [30].

A high level of RF energy has the potential to damage or destroy most of electronic-based systems identified earlier: radar, communications equipment, electrical power generators and engine controls, navigational aids, electronic flight controls, and weapon fuses.

Solutions require focus on equipment layer hardening. Additionally, wiring should be replaced with fiber optics, and wiring, housings, plugs and airframe materials should be effectively earthed and grounded. The cost of the onboard electronics represents a substantial proportion of an aircraft's total cost and the cost of maintaining significant holdings of spare parts may be prohibitive [31].

7.6.3 Safety and Arming Systems

Electroexplosive devices (EED) are found in a wide range of military systems, including warheads, rocket motors, gas generators, cable cutters, thermal batteries, and flares. They have the advantages of reliability, low-power requirements, and rapid response time. However, most EED are susceptible to uncommanded initiation and therefore need to be protected against conducted and radiated electromagnetic interference. Two main types of EED are in service: the bridgewire (BW) and the conductive composition (CC) [32].

Various organizations, such as the Australian Ordnance Council [33], have published minimum electromagnetic environments for which ordnance is designed to be safe. MIL-STD-464 specifies that an additional 6-dB margin should be provided for ordnance over that mandated for other equipment [34]. This means that the maximum range of RF DEW against ordnance is half that of an electronic system protected in accordance with the minimum requirements of MIL-STD-464.

7.6.4 Personnel

The main direct threat to personnel from microwave radiation is through heating of the body tissue [35]. The short duration of an RF DEW pulse is not thought to raise the body temperature sufficiently to pose a hazard to personnel [36, 18]. However, CW RF DEW do have the potential to be used in the antipersonnel role. The heating of body tissue has been known to cause cataracts, corneal opacities, testicular damage, lesions, hemorrhages, and induced fevers [37]. In some animals, a transient rise of 10 degrees Celsius in body temperature can be dangerous, and a sustained increase of 1 degree can be fatal [38].

Although pulsed RF DEW pose little direct threat to personnel, an indirect threat may exist in future land warfare due to the significant amount of electronic equipment carried by soldiers. In addition to radios, soldiers' personal equipment will include night-vision devices, hand-held computers, and navigation devices, all of which are vulnerable to RF DEW.

While personal equipment can be protected by means of the hardening techniques described earlier in this chapter, such measures are likely to increase their size and weight, making them more difficult to carry and employ. Because of this, the screening effect provided by the terrain surrounding ground forces may well provide the best protection for land forces.

7.6.5 Support Equipment

Land forces also employ a wide range of support equipment such as ground-based radar, generators, laser-range finders, water purifiers, and kitchens. Any of this equipment that relies on electronic devices will be vulnerable to RF DEW. Platform hardening may be difficult and could affect the operation of the equipment. The best defense measure against RF DEW may be terrain screening.

7.6.6 Command Posts

The establishment of an exclusion zone around a command post is likely to be easier than for an individual vehicle, simply due to the greater availability of manpower.

The best defense is the establishment of an exclusion zone around the command post, at least for vehicles, by the use of checkpoints and active patrolling. Defense against conventional explosives will require the exclusion of conventional forces from the command post itself, which will also aid defense against RF DEW. The establishment of the command post outside urban terrain may assist defense.

Other current tactics that will assist in protection against RF DEW include concealment of the command posts, since existing conventional weapons systems such as MLRS have similar concentrations to that concentration likely for RF DEW; dispersion in main and alternate command posts; and regular changes of location.

7.6.7 Logistics Installations

Major logistics installations are likely to be out of the range of artillery and tank guns. Considerations for the defense of smaller installations such as a brigade maintenance area will be similar to those applying to a command post. However, additional difficulty will be experienced in the exclusion of potential threats from the installation due to the volume of traffic entering and leaving the facility and the likely need to employ local civilian staff.

The establishment of an exclusion zone may be particularly difficult in situations in which large numbers of personnel (including possibly civilians of questionable loyalty) and civilian vehicles are working in close proximity to, or within, the facility to be protected. This problem is likely to arise in urban terrain, rear logistics areas, and U.N. operations.

7.7 Summary

RF DEW potentially pose a significant threat to land forces through the neutralization and destruction of their electronic systems. One of the major advantages offered by such weapons is the reduced requirement for accuracy compared to many conventional weapons, such as artillery. In order to protect land forces against the effects of RF DEW, all electronic systems should be protected both in accordance with the specifications of MIL-STD-464 and by suitable tactics and procedures to minimize the likelihood of illumination. Difficulties may be created by the size of the area surrounding headquarters or other installations from which RF DEW must be excluded.

For offensive operations, the impact of RF DEW against platforms protected in accordance with MIL-STD-464 may only be guaranteed for ranges up to approximately 100m or 1 km, depending on frequency. It is likely also that the impact can be greatly increased by taking advantage of those frequencies at which particular systems are most susceptible.

Endnotes

[1] "The Technology That Won the War," *Reinventing the Wheel,* BBC Radio 4. Available at: http://www.bbc.co.uk/education/archive/wheel/war.htm.

[2] O'Neill, J. J., *Prodigal Genius: The Life of Nikola Tesla,* New York: Ives Washburn, 1944, p. 239.

[3] Defense Intelligence Agency, *Soviet Military Power,* Chapter 3, 1987.

[4] "Russia Leads in 'Pulse' Weapons," *Jane's Defence Weekly,* Vol. 18, No. 15, October 10, 1992.

[5] *Army Science and Technology Plan,* §10, "Electronic Warfare/Directed Energy Weapons," U.S. Department of Defense, 1998.

[6] Fowler, C., et al., "Production of Very High Magnetic Fields by Implosion," *Journal of Applied Physics,* Vol. 3, No. 3, 1960, pp. 588–594.

[7] "Weapons Systems," *Jane's Defence Weekly,* Vol. 19, No. 24, June 12, 1993.

[8] Pasternak, D., "Weapons: The Pentagon's Quest for Non-Lethal Arms Is Amazing. But Is It Smart?" *US News and World Report,* 1997, pp. 38–46.

[9] FY 1997 Defense Technology Area Plan for Weapons, §3.9, U.S. Department of Defense, 1996.

[10] FY 1998 USAF Military Space RDDS, U.S. Department of Defense, 1997.

[11] FY 1997 Defense Technology Area Plan for Weapons, WE.22.09, U.S. Department of Defense, 1996.

[12] Joint Science and Technology Master Plan, WE.19.08, U.S. Department of Defense, 1996.

[13] "Contracts Awarded, R&D," *Jane's Defence Contracts,* February 1996.

[14] Kopp, C., "EMP—The Emerging Electromagnetic Threat," *Australian Aviation,* 1995, pp. 50–54. *See also* Author, "Disabling Technologies—A Critical Assessment," *International Defense Review,* Vol. 27, No. 7, July 1994.

[15] "Non-Lethal Microwave Weapons," *Defence Systems Daily,* June 14, 1999.

[16] Director of Central Intelligence Directive No. 6/1 SIGINT Committee, May 12, 1982.

[17] "Signals, TNO-PML Studies Pulse Weapon Integration for Future Frigates," *Jane's Navy International,* Vol. 101, No. 008, Oct. 1, 1996.

[18] Wood, L., Statement to House of Representatives Committee on National Security, Military Research and Development Subcommittee, on the threat posed by EMP to U.S. military systems and civil infrastructure, Washington, D.C., July 16, 1997.

[19] Trippe, A., "The Threat of Electromagnetic Pulse," *National Defense,* 1984, pp. 22–27.

[20] Sutton, P., "RF Directed Energy Weapons (DEWs)," *RMCS Army Staff Course Notes,* 1995.

[21] MIL-STD-188-125 is divided into two parts: MIL-STD-188-125-1, *High-Altitude Electromagnetic Pulse (HEMP) Protection for Ground-Based C4I Facilities Performing Critical, Time-Urgent Missions Part 1 Fixed Facilities,* U.S. Department of Defense, July 1998; and MIL-STD-188-125-2, *High-Altitude Electromagnetic Pulse (HEMP) Protection for Ground-Based C4I Facilities Performing Critical, Time-Urgent Missions Part 2 Transportable Systems,* U.S. Department of Defense, March 1999.

[22] *Electromagnetic Pulse (EMP) and TEMPEST Protection for Facilities,* Engineering and Design Pamphlet EP 1110-3-2, U.S. Army Corps of Engineers, December 1990.

[23] MIL-STD-464, *Electromagnetic Environmental Effects—Requirements for Systems,* U.S. Department of Defense, March 1997.

[24] Jump, M., and W. Emberson, "Ship Electromagnetic Pulse Survivability Trials," *Naval Engineers Journal,* May, 1991, pp. 136–140.

[25] Neuheuser, H., "The Nuclear Electromagnetic Pulse," *Military Technology.* 1985, pp. 98–100.

[26] Cabeyan, H. S., "Current Status of Higher Power Microwave Effects and Simulation," *Lawrence Livermore National Laboratory,* December 1986.

[27] While generators with power output greater than 10 GW are theoretically possible, they pose a number of practical problems that make them unlikely. The atmospheric ionization resulting from the use of such powers results in high power losses and poses a significant threat to the delivery platform that may be unacceptable for nonexplosive RF DEW.

[28] MIL-STD-464, *Electromagnetic Environmental Effects—Requirements for Systems,* U.S. Department of Defense, March 1997, Table 1C, p. 7.

[29] Kopp, C., *An Introduction to the Technical and Operational Aspects of the Electromagnetic Bomb,* Fairbairn ACT: Air Power Studies Centre, Australia, 1996. *See also* Kopp, C., *The E-Bomb: A Weapon of Electrical Mass Destruction,* Clayton, Melbourne, Vic.: Monash University, 1996.

[30] Soper, G., and K. Casey, "Understanding the EMP Threat," *Defense Electronics,* 1987, pp. 156–169.

[31] Kopp, C., "A Doctrine for the Use of Electromagnetic Pulse Bomb," Fairbairn ACT: Air Power Studies Centre, Australia, July 1993.

[32] Nott, A., and J. Whitelaw, "Electromagnetic Radiation Hazard Testing of Electroexplosives in Australia," *Journal of Battlefield Technology,* Vol. 3, No. 2, March 2000.

[33] Australian Ordnance Council, "Guidelines for the Preclusion of Electroexplosive Hazards in the Electromagnetic Environment," *AOC Pillar Proceedings,* October 1994, pp. 236–294.

[34] MIL-STD-464, *Electromagnetic Environmental Effects—Requirements for Systems,* U.S. Department of Defense, March 1997, p. 5.

[35] Mumford, W., "Some Technical Aspects of Microwave Radiation Hazards," *Proc. IRE,* Vol. 49, February 1961, pp. 427–447.

[36] *EMP Engineering and Design Principles,* Bell Telephone Laboratories, 1975, p. 138.

[37] See, for example, Moore, W., *Biological Aspects of Microwave Radiation: A Review of Hazards,* U.S. Department of Health, Education and Welfare, Public Health Service, National Center for Radiological Health, July 1968; Leary, F., "Researching Microwave Health Hazards," *Electronics,* 1959, p. 49; and Weiss, M., and W. Mumford, "Microwave Radiation Hazards," *Health Physic,* Vol. 5, June 1961, pp. 160–168.

[38] Solait, O., and H. Schwan, "Techniques for Relative Absorption Cross Section Determinations," in *3rd Annual Tri-Service Conference on Bio-Effects of Microwave Radiating Equipments, RADC-TR-59-140,* 1959. See also Mumford, W., "Some Technical Aspects of Microwave Radiation Hazards," *Proc. IRE,* Vol. 49, February 1961, pp. 427–447.

8

Electronic Warfare and Digitization

8.1 Introduction

The defining feature of EW has always been the electromagnetic spectrum. Traditional EW activities have been aimed at degrading an adversary's ability to use the electromagnetic spectrum or protecting friendly use of the spectrum. Expressed in the language of the OSI model for communications systems, traditional EW has mostly been a physical-layer activity. This means that it has been oriented mostly at electromagnetic systems. More recent technology has led to the development of electromagnetic weapons that can disrupt and damage a range of systems other than those not designed to be emitters or receivers of electromagnetic radiation.

The key change in the nature of EW on a future digitized battlefield will be its orientation toward the *network,* leading to a proliferation of opportunities for EW. While many of the EW techniques in the future will be the same as the ones currently used, the focus will change from being primarily a physical layer to focusing on attacks on network security and the security services that protect against these attacks. This same network technology will also greatly increase the capability of friendly EW.

This chapter looks at the future of EW and its targets, examining a number of key areas of communications technology, with likely advances and their impact on EW. These key areas are networking, ultra-wideband radio, improved high-frequency radio technology, software radio, quantum computing, and quantum cryptography.

8.2 Network Issues

In the foreseeable future, the most significant change likely to occur in the targets for EW due to battlefield digitization is the evolution from a collection of related, but separate, communications systems to a network. This network will facilitate the passage of information between any two points on the battlefield, as well as between any point on the battlefield and terminals associated with other networks, such as joint and even multinational systems.

In Chapter 2, we outlined the motivations for migrating to a network architecture: to maximize the effectiveness of the passage of information between sensors, command elements, and weapons systems. For the reasons outlined in Chapter 2, primarily related to a tradeoff between mobility, capacity, and range in communications links, it is unlikely in the near future that this network can be provided with a single communications technology. In other words, it is unlikely that an evolution of the equipment and protocols associated with the CNR subsystem will be able to meet all the requirements of the tactical communications system; the same applies to the trunk and tactical data distribution subsystems. Furthermore, there is no single technology on the horizon that could replace all of these systems.

While a single homogeneous network is not likely, a single logical network is both desirable and achievable using current and developing technology. The major changes that will occur in this evolution to a network are seamless integration of all subsystems, the provision of truly mobile networking, and the use of ad hoc network technology. Seamless integration of all subsystems will enable the passage of information between any two points on the battlefield. Mobile networking technology will allow stations to move at will through the network without being constrained in their location. Ad hoc network technology will allow the network to be self-forming, without the need for large numbers of dedicated base stations throughout the area of operations.

8.2.1 Seamless Integration

Given the lack of a suitable technology from which to build a homogeneous tactical communications network, seamless integration between a number of subsystems is the only means of providing a single logical network. While this integration has not occurred in the tactical communications system, it is becoming increasingly common in commercial systems. Examples include the global telephone network, containing interfaces for both fixed telephones (both analog and digital) and mobile telephones. The telephone network has also evolved to carry data as well as voice. Another example is the Internet, whose terminals are connected by a range of interfaces, including high-performance local area networks (LANs) providing capacities of 10 Mbps or more, cable modems (approximately 0.5 Mbps), dial-up modems (up to 56 Kbps), and mobile dial-up connections (up to approximately 10 Kbps).

The practical implementation of a single logical network requires the use of common protocols, especially at the network layer and higher layers, the use of suitable link-specific protocols at the physical layer and data-link layer, and ubiquitous encryption to provide security for the network. Given the current commercial technology, it is most likely that the majority of information-processing equipment that will use future tactical communications systems will be based on ruggedized computers using the same TCP/IP protocol used in the Internet.

An outline view of a typical modern network is illustrated in Figure 8.1. It is based on a hierarchical structure. Each terminal is connected to a local network. Each local network is connected to one or more other networks via a router, denoted "R" in Figure 8.1. The purpose of the router is to route data between the various local networks. A group of networks and routers forms an *autonomous system* (AS). In a tactical network, a local network may be the internal network of a headquarters, while an autonomous system may consist of all the networks and routers in a formation.

One of the advantages of the hierarchical structure shown in Figure 8.1 is that it minimizes the requirement for network terminals to understand

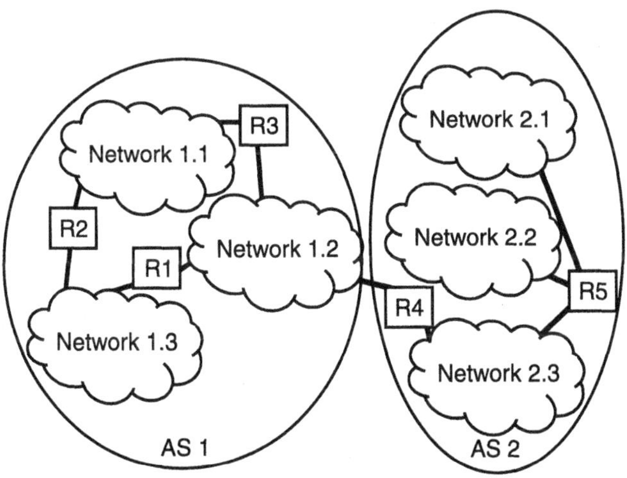

Figure 8.1 Outline structure of a TCP/IP network.

the structure of the network. Terminals only needs two pieces of knowledge about the network: the identities of the other terminals attached to their local network (to whom they can therefore transmit data directly) and the address of the router to which all other traffic should be sent. Routers require knowledge of the next hop to route data between terminals with their local AS and the identity of the router that handles traffic destined for outside the AS. Only these boundary routers require explicit knowledge of the outside network, and even here the knowledge required relates only to the first hop outside the AS.

Seamless integration will provide the network for network-centric warfare, underwritten by a variety of technologies, including those associated with evolutions of cellular mobile telephone systems (especially third-generation systems), satellite technology, network protocols, and miniaturization of electronics. The move to network-centric warfare will lead to a proliferation in transmitters, each of which is a potential target for ES, and even EA.

This organic, tactical network will be supported by a range of overlaid communications systems, including operational and strategic-level military systems. The U.S. *global information grid* (GIG) is one such concept, aiming to provide seamless integration throughout a reliable, assured, cost-effective, global network. An important means for providing this level of connectivity will be the incorporation of multiple layers of airborne rebroadcast using aircraft, UAVs, and satellites [1].

8.2.2 Mobile Networks

Users in the tactical communications system should be able to move from one part of the network to another, and transparently receive the same services in their new location. Depending on the equipment used, this roaming may be provided using a wireless connection or require connection to a wired network at the new location. Mobile networking protocols, developed in recent years, provide this service using a forwarding agent (Figure 8.2). When station *B* moves from network 1.3 to network 1.2, it changes its address to a value lying in network 1.2. One station (labeled "F") in network 1.3 acts as a forwarding agent, receiving any data addressed to *B* and forwarding it to *B* at its network 1.2 address. So long as a forwarding agent exists for each local network, stations can roam at will through the network. The cost of providing mobility is some double-handling of data destined for a roaming station in passing through the forwarding agent.

The implementation of mobility, particularly wireless mobility, has implications for the management of encryption keys. It will be necessary to provide either a common key for use in affiliation across the whole network

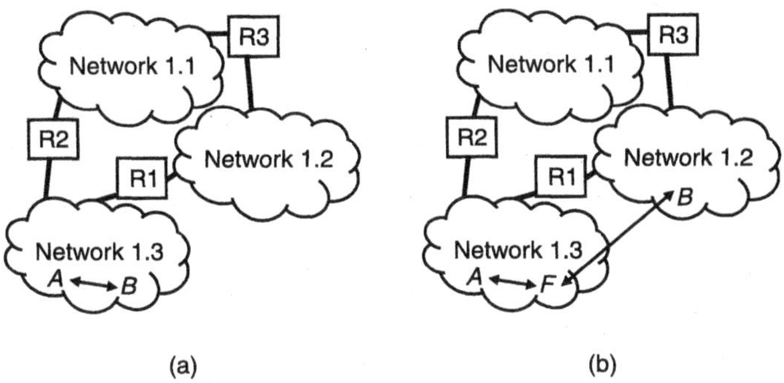

Figure 8.2 Mobile networking. Station *B* is (a) in its home network, and (b) roaming in a different network.

or to provide mobile users keys for use in different parts of the network, imposing difficulty in guaranteeing the security of such widely distributed keys. The use of wired network connections may reduce difficulties with security by allowing individual stations to connect to the network without the use of encryption systems.

In areas exposed to an adversary EA threat, it is likely that the capacity of wide-area, fully mobile, tactical wireless communications systems, such as CNR or developments on it, will remain limited. Nonetheless, there is the potential for local-area communications systems, based on technologies such as Bluetooth [2], to offer high-capacity, truly mobile communications. These systems may offer ranges of no more than tens of meters, operate in parts of the electromagnetic spectrum that are not regulated (overcoming the need to take spectrum away from other tactical uses), and offer data rates up to 2 Mbps. They are likely to employ a range of EP techniques, including FH [3], to reduce their susceptibility to both natural and man-made interference. The combination of short range and the use of EP will enable communications with very low transmit powers, maximizing battery life [4].

The provision of such a local-area communications system will enable the networking of the sensors, weapons, and communications systems carried by an individual soldier without the weight and inflexibility of connecting cables. It will also enable the networking of small groups of soldiers, providing a "section LAN" on which data from sensors and weapons can be shared.

8.2.3 Ad Hoc Networks

When we speak of mobile communications in current, commercial networks, we mean only that the user terminal is mobile. The network itself is very much fixed in place. In a cellular telephone system, for example, all communications passes from a mobile handset to a base station, which is in turn connected to the fixed network. In tactical communications systems, not only is the user mobile, but also the whole network must be able to move with the force it supports, and adapt its structure to changes in the disposition of forces as required.

In an *ad hoc network,* stations cooperate to build the network. Stations communicate using a common wireless channel. Each station can communicate directly with one or more of the other stations in the network, but it is unlikely that any one station can communicate directly with all of the other stations. Stations on the network are therefore required to act as relays. Data is carried through from source to destination by being passed from one relay to the next. Each station maintains a list of the stations to which it can directly communicate. Connectivity information is built up and distributed by each station [5].

In the example network shown in Figure 8.3, B can communicate directly with A and G. B may send data to E via the path BGE or the path BACE. Each of these paths would have an associated cost, which may be as simple as the number of hops involved. B would choose the least cost path, transmitting the data over the first hop. The relay station (e.g., G) then transmits the data over the next hop, with this process continuing until the data reaches its destination.

In larger ad hoc networks, stations may form themselves into clusters. A small number of stations may then take on the role of communicating between clusters, possibly using higher transmission power to do so. The forming of clusters helps to maximize frequency and battery life reuse by minimizing transmission power. In the example in Figure 8.4, E and F have taken on the role of intercluster communication.

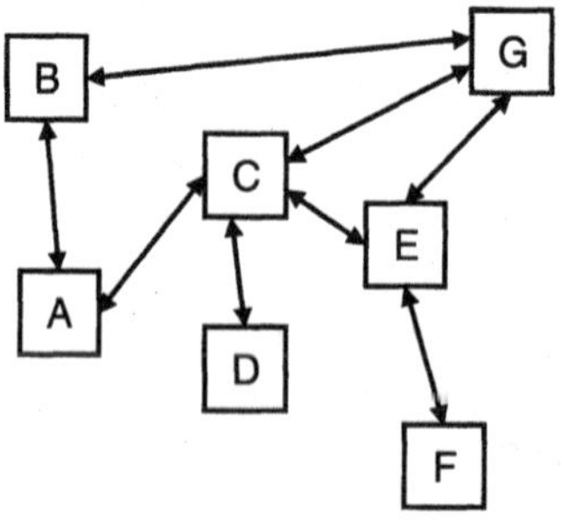

Figure 8.3 Example of connectivity within an ad hoc network.

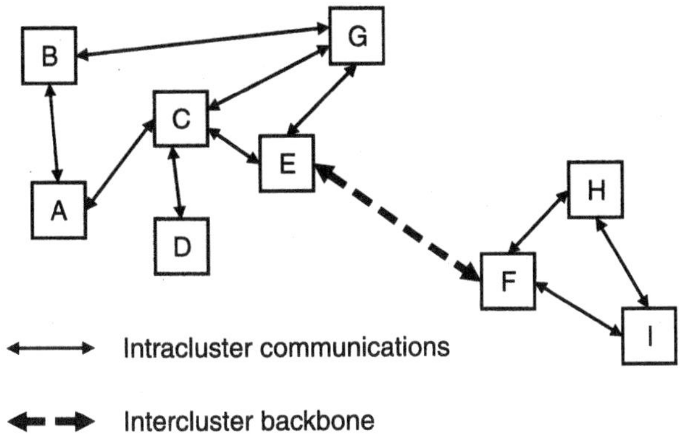

Figure 8.4 Example of clustering in an ad hoc network.

An ad hoc network may be integrated with a wider network by one of the stations on the ad hoc network acting as a gateway.

The major utility of an ad hoc network in the tactical communications system is the fact that the network is formed by the terminals, without the requirement for a specific infrastructure to be deployed.

8.2.4 Implications for EW

The advent of the battlefield network will bring with it a number of characteristics already found in commercial networks. One of the most important of these in the context of EW is the concept of *security services* [6]. These are a generalization of the use of encryption to protect information against unauthorized access. The security services are confidentiality, authentication, integrity, nonrepudiation, access control, and availability.

Confidentiality Information transmitted through the network should be available for reading only by authorized parties. This service is traditionally provided in military systems by encryption. Generalizing the subdivision of encryption from Chapter 5, the confidentiality service may conceal the contents of the message; the contents of the message and header information, such as the identities of sender and receiver; or the very existence of the message, as is provided by the bulk encryption used in the trunk subsystem or by LPI techniques such as spread-spectrum communications (in which case confidentiality can be seen as protecting the location of the transmitter) [7].

Authentication Each party to an exchange of information across the network should be able to guarantee the identity of other parties involved. This applies to both senders and recipients of information.

Integrity Information transmitted though the network should be protected against modification by an adversary.

Nonrepudiation The sender of a message should receive a receipt that guarantees that the message was received by the intended recipient, preventing the recipient from later denying receiving the message. Similarly, a recipient of a message should receive an attachment to the message that can be used to prove that the message was sent by a particular party. One impact of nonrepudiation is to protect against counterfeit information being inserted into the network. Nonrepudiation cannot exist without authentication.

Access Control Access to systems connected to the network should be limited to authorized parties. Access control includes physical security as well as electronic measures, such as the use of passwords.

Availability The capacity of the communications system should be protected, preventing an adversary from degrading system performance.

The exploitation of an adversary's network, whether by using ES or EA, can be seen as an attack against one of more of these security services. Possible attacks include the following:

- *Interception.* An unauthorized party may attempt to gain access to data transmitted across the network, or to a portion of this data, such as its external characteristics. Interception is an attack on confidentiality and possibly also on authentication, and encompasses all of the aspects of ES discussed in Chapter 3 [8].
- *Modification.* Following interception, an unauthorized party may modify this data and reinsert it into the network. Modification is an attack against integrity.
- *Fabrication.* An unauthorized party may insert counterfeit information into the network. Protection against fabrication is provided by authentication and nonrepudiation.
- *Interruption.* Also known as a *denial-of-service attack,* interruption aims to make become unavailable or unusable. It is an attack on availability.

The taxonomy of these attacks is based on what the attacker is trying to achieve, which means that there is not a simple one-to-one correspondence between the types of attack and the security services used to protect against them.

The division of EW into ES, EA, and EP still makes sense in the context of the network. However, they should be seen in the context of the security services that they are aiming to degrade or provide, rather than purely in terms of their relationship to the electromagnetic spectrum. ES is the means of exploiting an adversary's use of the electromagnetic spectrum using only passive systems (i.e., receivers). In the language of security services, ES is about interception, that is, an attack on confidentiality (in the broad sense defined previously). EA is the means of exploiting an adversary's use of the electromagnetic spectrum using active means (i.e., transmitters). Defined in terms of the desired outcome on the adversary's network, these attacks may take the form of modification, fabrication, or interruption. When applied to communications and information systems, EP is the provision of security services to protect friendly capabilities from the effects of friendly EA, and adversary ES and EA.

The traditional subdivision of ES into search, intercept, DF, and analysis remains valid. Details of the equipment will change to enable, for example, an intercept receiver to monitor digital network traffic. ES is primarily used as an attack on confidentiality, whether of the contents of a message, the external characteristics of a message, or the location from which it is transmitted.

Similarly, even though the aims of EP may be reframed in terms of network security services, the basic techniques for providing LPI and resistance to jamming will not change.

The application of EA in the context of security services and the associated attacks can be understood in terms of the mission given to an EA asset. This mission will include a task (e.g., to jam the adversary command net) and a required outcome (to deny communications.) The outcome specified here is interruption. An outcome "in order to force the net to operate in plain" specifies an interruption attack, leaving the adversary vulnerable to a later interception attack. EA can be used to provide modification, fabrication, and interruption attacks. Jamming and neutralization are exclusively associated with interruption; electronic deception may be associated with all three.

The use of wireless networking protocols creates new vulnerabilities, making interruption possible by jamming and deception. Transmitting signals that imitate the transmissions of an adversary's data communications systems, especially for protocols based on carrier-sense multiple access (CSMA), may trick the adversary's systems into thinking that a channel is active, and prevent them from attempting to transmit. Hence, interruption may sometimes be achieved at much lower powers than are required for jamming. This opens up a new class of electronic deception, aimed principally at the adversary's network rather than at the adversary commander. The detailed coordination of this type of electronic deception is probably more closely tied to jamming rather than the force's deception plan.

New vulnerabilities will also be created by the use of ubiquitous wireless networking of sensors and weapons systems, increasing the potential impact of electronic minefields. These wireless networks will add an electromagnetic dimension to the signature of the smallest groupings of soldiers wherever they operate.

The digitization of the battlefield will cause the number of targets available for EW to increase significantly. This will place a greater strain on the already scarce EW assets, especially on ES. Depending on the strength of the algorithms used, the use of encryption throughout the network may reduce the need for interception if the algorithms are strong. If the algorithms are susceptible to cryptanalysis, network-centric warfare will facilitate the coordination of collection and processing or the intercepted traffic.

Universal encryption will increase the difficulty of obtaining internal information from intercepted transmissions. The use of network encryption keys, rather than separate keys for individual links or nets, however, may also introduce new vulnerabilities. The larger the volume of data that is transferred using a key, the more vulnerable that key is to cryptanalysis. The interception of preambles used in the affiliation of mobile stations to networks, and their retransmission in other parts of the network, makes possible the use of electronic deception to carry out interruption attacks. The value of encryption keys stored in captured equipment may also be increased, potentially allowing that equipment to be used in a wide range of deception attacks. The extensive use of ad hoc networks potentially increases this vulnerability. One method for overcoming the vulnerabilities created by the use of preambles is to employ an alternative means of synchronization for encryption and spread-spectrum communications. The use of a common time reference, which may be derived from GPS, is one possibility.

The extensive use of commercial off-the-shelf (COTS) equipment in modern tactical communications systems is also a source of increased vulnerability. Much of this equipment does not conform to military standards for EP, such as TEMPEST and MIL-STD-464. Furthermore, commercial wireless network protocols are not designed to operate in a hostile electromagnetic environment, and are vulnerable to a variety of attacks, especially interruption. In general, COTS equipment will be more vulnerable to jamming, deception, and neutralization.

As well as providing new targets, network-centric warfare will transform the planning and coordination of EW itself. It will enable better coordination of EA as a fire and more effective use of ES assets in their combined roles of collection and providing steerage for EA. In this context, ES assets are simply sensors and EA assets are weapons platforms.

8.3 UWB Radio

Ultra wideband (UWB) radio, also known as *impulse radio,* transmits information in a sequence of short pulses, typically between 0.1 and 1.5 ns. As illustrated in Figure 8.5, the information content is encoded in time, rather than frequency or amplitude. In its simplest form, a one may be encoded as a pulse arriving shortly before a nominated time, a zero as the pulse arriving shortly after this time [9].

While a uniform pulse-train spacing may be used (i.e., $T_j = jT_1$), multiple access is best supported by a system incorporating a pseudorandom pulse-train spacing, sometimes referred to as *time-hopping.* The use of a pseudorandom pulse-train spacing (i.e., the sequence $\{T_j\}$ chosen to be pseudorandom) prevents the loss of a large number of consecutive bits due to inadvertent synchronization between two transmitters. The use of a pseudorandom pulse-train to provide multiple access for two transmitters is illustrated in Figure 8.6. The variation in the pulse-train timing prevents regular collisions between transmitters, but guarantees that some clashes will occur. Unlike CSMA, collisions that cause loss of data in impulse radio are

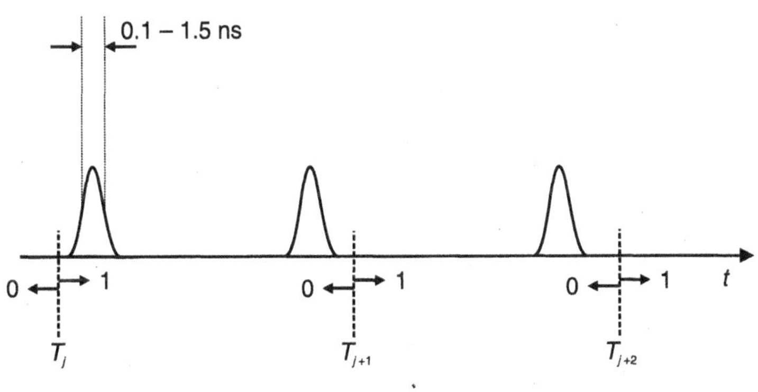

Figure 8.5 Time-coding of information in UWB radio.

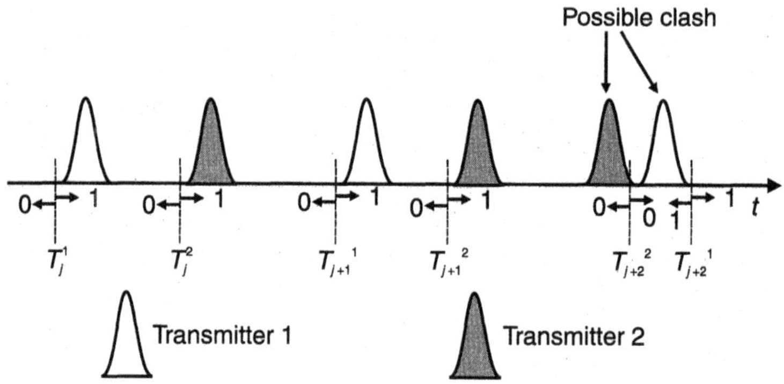

Figure 8.6 Multiple access in impulse radio.

not primarily due to overtransmission, but arise where the receiver receives two impulses indicating conflicting values for that symbol. The near-far effect prevents the design of a multiple-transmitter, multiple-receiver UWB radio system that uses synchronization of transmitters to overcome collisions.

The use of baseband pulse modulation enables impulse radio to have an extremely wide bandwidth, typically in excess of 25% of the center frequency of the signal. The relationship between pulse length and the frequency content of transmissions is illustrated in Figure 8.7. The transmission of information using time-modulation of short pulses is a form of spread-spectrum communications. Spreading gains of 45 dB (30,000) have been achieved in prototype systems [10]. Like the other forms of spread-spectrum communications discussed in Chapter 5, this frequency spreading can also provide LPI.

The use of baseband transmission creates the possibility of building extremely simple receivers, without the requirement for a conventional RF front end. These receivers have the potential of having a much lower cost than traditional receivers based on FDMA.

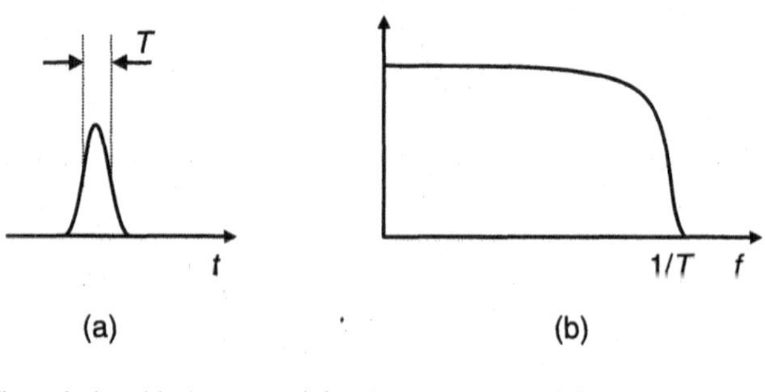

Figure 8.7 The relationship between (a) pulse length and (b) frequency content.

Very low transmission power levels can be used. Once a receiver is synchronized to a transmission, it does not need to detect pulses. In the example shown in Figure 8.5, the receiver has only to decide whether it is more likely that the pulse arrived before or after the nominated time. This enables the use of low transmission powers for the impulses. The average power is further greatly reduced by the fact that the transmitter is active for only a very small proportion of the time. Average transmitter powers are expected to be on the order of 1 mW. The initial synchronization can be achieved by the transmission of a long synchronization sequence, or by the transmission of a short, higher-power synchronization sequence.

Communications applications for UWB radio include short-range, high-capacity communications systems. It is also possible to construct positioning systems with very high accuracy (better than 1m) and radar imaging applications capable of operating through walls and other obstructions.

UWB radio has the potential to provide high-capacity, mobile communications within small areas, essentially removing the requirement for frequency management. Multiple access is achieved by assigning a different time-hop sequence to each user community. There remain significant challenges in establishing the feasibility of providing efficient multiple access using impulse radio and in preventing interference with other communications and navigation systems. For example, current systems are believed to cause interference with GPS to a range of approximately 30m [11]. The design of antennas that provide efficient operation with very high bandwidth is also an open problem. Additionally, current regulations for frequency management in most countries explicitly prohibit the use of broadband transmitters that operate across bands allocated for other purposes. These regulations tend to be particularly strict for frequencies allocated for safety and emergency uses. In addition to the technical challenges, therefore, there is a requirement for the reframing of regulations to accommodate UWB radio [12].

8.3.1 Implications for EW

The use of UWB radio will make detection, intercept, and DF impossible using most conventional ES equipment. Narrowband receivers will see the UWB transmissions as low-power noise, while wideband systems will see these transmissions as very short pulses, which they are not designed to receive.

The task of a search receiver for UWB radio is of a similar level of difficulty to other spread-spectrum transmissions. The difficulty is increased, however, by the very high spreading gains, which can be in excess of 40 dB.

Without knowledge of the pseudorandom pulse-train spacing employed, the search receiver must actually detect the pulses. This is likely to be extremely difficult to do reliably, given the very low power levels used.

The difficulty of design of a UWB intercept receiver depends on whether or not the pseudorandom pulse-train spacing is known. If this spacing is known, the intercept receiver is constructed in the same way as a communications receiver. If the spacing is not known, the intercept receiver may be able to record the sequence of pulse interarrival times. In this case, reliable detection of pulses is of key importance.

DF of a UWB transmission is hindered by its very short duration. It may be possible to design a DF system that measures the time difference of arrival of pulses at two or more antennas, and uses this information to infer a bearing to the transmitter. Conventional techniques, such as Watson-Watt or Doppler DF, are unlikely to be useable, as they rely on the length of a transmission being much greater than those used in UWB radio.

Difficulties with search, intercept, and DF will be increased by the presence of multiple UWB transmitters operating simultaneously.

Jamming of a UWB transmission is hindered by the very high processing gains achieved by these radios. Efficient operation of a large number of UWB transmitters in one area and minimization of interference with other users of the electromagnetic spectrum require the use of very low average powers on the order of 1 mW. Conventional man-portable CNR transmitters, however, operate at transmit powers of up to 1 to 10W. In other words, in comparing jamming of conventional CNR to UWB radio, the first 30 to 40 dB of spreading gain of a UWB transmission simply overcome the transmitter power difference. The effect of a UWB transmitter increasing its transmit power to overcome jamming will be to increase its interference caused to all conventional transmissions in the same area and to increase the probability of intercept and detection.

The jamming waveform used against a UWB transmission could be either noise jamming from a barrage jammer or, more likely, a sequence of short pulses with bandwidth similar to the transmission being jammed. The very high synchronization accuracy required between transmitter and receiver may make pulsed jamming attractive, in which the jammer transmits a sequence of pulses long enough to cause the UWB receiver to lose synchronization.

The use of electronic deception against an adversary communications system based on UWB radio relies on having knowledge of the pseudorandom pulse-train spacing used by that network. This is equivalent to knowing the key in a conventional, secret-key, secure communications system. Electronic deception is therefore likely to be possible only under exceptional circumstances.

8.4 HF Radio

In many armies, the use of satellite communications systems has led to a reduced emphasis on HF radio as a beyond-line-of-sight means of communications. While satellite infrastructure is expensive, it enables the principal limitations of conventional HF systems (i.e., capacity and reliability) to be overcome. In conventional HF systems, raw data capacity has been limited to approximately 2.4 Kbps for a 3-kHz channel. Reliability is limited by the requirement to use sky-wave propagation to achieve long distances, which imposes a requirement for a high level of operator expertise, with a requirement for ongoing training.

Recent advances in HF radio technology, leading to improved data rates and reliability, have given new life to HF communications, reinforcing the need for complementary EW systems. Applications include providing an alternate means to satellite for long-distance communications and long-range, special-forces communications.

8.4.1 High-Speed HF Modem Technology

Conventional HF modems, using QPSK modulation, achieve raw data rates of approximately 2.4 Kbps over a 3-kHz channel. Recent advances in modulation techniques, employing varieties of quadrature amplitude modulation (QAM), enable data rates up to approximately 9.6 Kbps, depending on the quality of the link. Some systems also attempt to measure link quality on a regular basis, selecting the maximum data rate that can be achieved. This change in data rate may be performed automatically or may require operator intervention.

8.4.2 HF Automatic Link Establishment

As long as a channel with sufficiently high signal-to-noise ratio is available, recent improvements in modulation technology enable useful raw data rates to be achieved using sky-wave communications. The engineering of a channel of sufficient quality has always required skilled operators who can adjust antenna characteristics and change operating frequencies to take into account slow changes in the characteristics of the channel. However, even skilled operators can have great difficulty working with short-term variations in the characteristics of the channel.

Automatic link establishment (ALE) describes a set of automatic procedures that can be used to establish and maintain HF sky-wave communications. It is usual that a station initiating a transfer of data will have available a number of channels. The connection begins by the initiating station sending a short message to permit receiving stations to measure link quality. This process is known as *sounding*. The receiving stations may measure link quality by means of the bit error rate or the channel signal-to-noise ratio. Information on channel quality is returned to the initiating station. The initiating station may then proceed to the transfer of data if the link quality is acceptable. Otherwise, it may return to sounding, using another channel. The time taken for link establishment depends on the number of channels that must be tried before one with acceptable quality is found, but is usually on the order of 5 to 20s.

Stations may carry out sounding to maintain current information on the quality of the available channels, even when they have no data to send.

ALE may be used for voice or data communications, and a number of standards have been developed [13]. Some data communications systems limit the maximum length of message that can be transferred to approximately one minute. This maximizes the ability of the system to take advantage of short-term variations in the channel to achieve high data rates, and reduces the likelihood of the channel quality becoming unacceptable during message transfer. In a hub-spoke network, limiting the maximum transmit time may also be useful as a means of providing fair access the channel.

The use of ALE makes possible long-range, reliable HF sky-wave communications without the use of skilled operators. However, ALE does impose overheads, which are necessary for evaluating channel quality and exchanging this information between stations.

8.4.3 Implications for EW

Siting of ES and EA assets to work against HF sky-wave communications is inherently more difficult than those working against line-of-sight communications. The main reason for this is that often there will not exist a location from which all transmitters can be received reliably, or from which all receivers could be jammed. This is illustrated in Figure 8.8. The likelihood of this situation occurring is increased by the use of directional antennas by one or more stations.

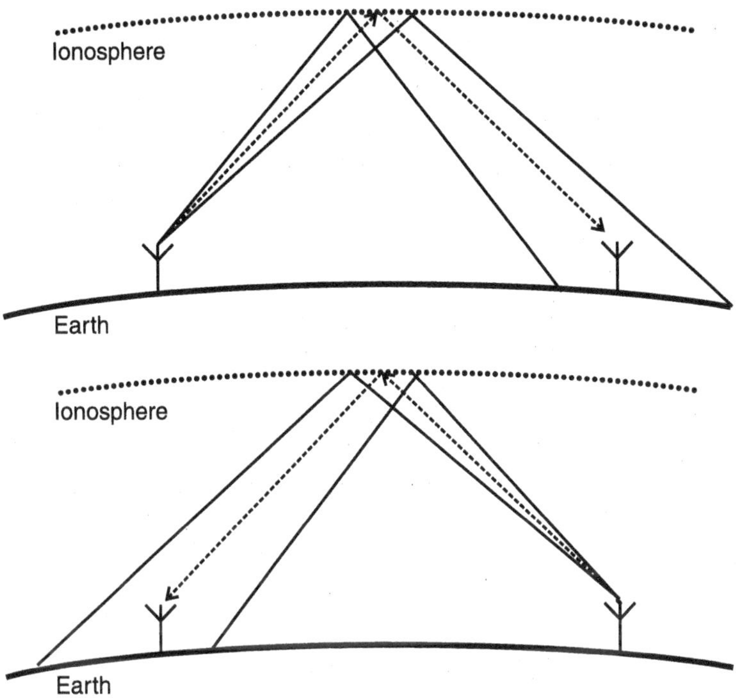

Figure 8.8 Areas may not exist where receiving of two communicating HF transmitters is possible.

This difficulty of siting EW systems is increased by the use of ALE. Without knowing all possible operating frequencies in advance, and without the possibility of obtaining regular soundings on these frequencies between the locations of EW equipment and the adversary's communications equipment, the performance of both ES and EA is likely to be unreliable.

The use of ALE also potentially creates opportunities for ES and EA. ALE is particularly vulnerable to jamming during link establishment. The successful jamming of the link establishment exchange prevents the transfer of data, and can probably be achieved by jamming either one of the stations involved. Even if the transfer of data is protected by active EP techniques such as DSSS, it is likely that a part, if not all, of the data exchanged for link establishment would not be protected. For a target system that operates in a hub-spoke configuration, where all communications is either to or from one station, ES and EA planning should therefore concentrate on the hub.

ALE systems that carry out regular sounding, even when they have no traffic to send, have increased vulnerability to ES.

If a suitable site can be found for a jammer, it may be able to force a reduction in channel data rate by reducing the channel signal-to-noise ratio, even if sufficient power is not available to prevent communications. If an automated HF data communications system performs this reduction in data rate automatically, the system's operator may have difficulty establishing whether the reduction in link quality comes from natural causes or jamming.

8.5 Software Radio

An ideal software radio is a multiband, multimode radio with a dynamic capability defined entirely in software in all layers of the protocol stack, including the physical layer [14]. This ideal radio allows such features as the air interface (including modulation technique, data rate, and channel bandwidth), voice coding, and encryption to be reprogrammed, potentially over the air. A simplified architecture for a transmitter that meets this ideal is shown in Figure 8.9, with a corresponding receiver architecture shown in Figure 8.10. In the transmitter, only two functions are performed after the digital-to-analog conversion: up-conversion to the transmission frequency and power amplification. Likewise in the receiver, analog processing is used only where it is absolutely required, mostly in the initial RF amplification and band-limiting to prevent aliasing in the analog-to-digital conversion. Practical systems may compromise on software programmability, most likely because of the limited power of available signal-processing technology.

The use of software radio technology will see some convergence of equipment in the various subsystems of the tactical communications system,

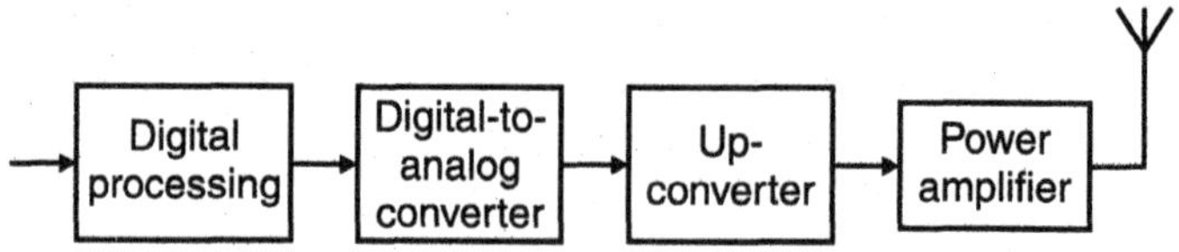

Figure 8.9 Simplified software radio transmitter architecture.

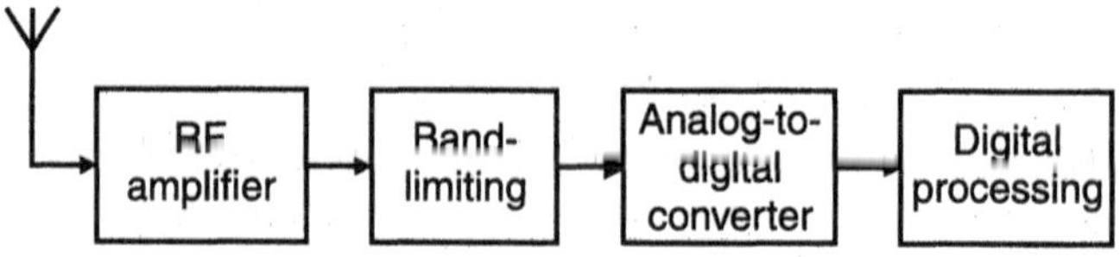

Figure 8.10 Simplified software radio receiver architecture.

especially between the trunk and CNR subsystems. However, in comparison with the CNR subsystem, the trunk subsystem will continue to be characterized by longer ranges, favoring the use of elevated, directional antennas and higher transmit powers, preserving the traditional tradeoff between capacity, range, and mobility.

8.5.1 Key Software Radio Technologies

There are a number of key technologies requiring development for software radio, including antennas, receiver RF processing and down-conversion, analog-to-digital conversion, signal-processing technology, and general-purpose processors.

Antennas One of the major aims for software radio is the construction of multiband radios. For the tactical communications system, efficient operation across the HF, VHF, and UHF bands is desirable, using a single antenna. It is also desirable that a single reconfigurable antenna be used for the various applications, allowing control (exercised by an operator or automatic controller) over such parameters as directivity and null steering. Such antennas are likely to be based on arrays, possibly containing thousands of elements.

Receiver RF Processing and Down-Conversion Because little or none of the selectivity of the software radio is contained in its RF stage, the requirement for low distortion is greater than for conventional radios, especially for tactical systems that must operate in a hostile electromagnetic environment. Any distortion introduced in this stage will cause leakage of power from one channel to another, possibly allow a narrowband jammer to jam signals in many channels, and leave a receiver vulnerable to inadvertent jamming from closely located friendly transmitters.

Analog-to-Digital Conversion The use of software radio systems in cellular telephone systems has led to a significant improvement in the speed and precision of analog-to-digital converters. Very low distortion and high precision are required in analog-to-digital converters in order to avoid leaving software radios vulnerable to off-channel jamming. A receiver's analog front end and its analog-to-digital converter are possibly the most critical parts of the system for operation in a hostile electromagnetic environment.

Signal-Processing Technology Recent years have seen significant improvements in the speed and precision of signal processing hardware, based on field-programmable gate arrays (FPGA) or application-specific integrated circuits, and software, based on programmable digital signal processors. Further gains are required in transmitter and receiver architectures and in the power of both hardware and software to make a truly programmable radio possible, into which almost completely arbitrary new waveforms can be introduced via software updates.

General-Purpose Processors To reach their full potential, software radios must implement not only physical layer, such as modulation, but complete protocols stacks. Implementation of higher layers is best performed on a high-performance, general-purpose processor.

The most ambitious tactical software radio project is the U.S. *Joint Tactical Radio System* (JTRS) [15], which aims to develop a family of software radios based on a common architecture, providing a range of services including voice, video, and data, and operating over a frequency range from 2 MHz to 2 GHz. European projects include the multirole multiband radio—advanced demonstrator model (MMR-ADM) funded by the German and French governments, and the programmable digital radio (PDR) funded by the United Kingdom's MoD.

Commercial applications are also under investigation for third-generation cellular telephone systems.

8.5.2 Implications for EW

Extensive deployment of software radio will create both opportunities and difficulties for EW. The use of COTS software radio technology is likely to increase vulnerability to jamming, with cost pressures limiting the quality of the analog front ends of receivers and their analog-to-digital converters. The ability to change a transmitter or receiver's EP, modulation scheme, data rate, and channel bandwidth in software will demand corresponding flexibility in ES and EA systems.

The flexibility offered by software radio will also reduce the cost of some EP techniques, such as FH and other spread-spectrum techniques, increasing their use in the tactical communications system. Responsive jammers capable of following a hopper will also be much easier to build, although the fundamental limitations of range discussed in Chapter 5 will still apply.

ES receivers based on software radio will feature high levels of flexibility. They offer the potential for automated, search, intercept, and DF of short-term and fast-changing signals, such as those generated using EP techniques including FH and burst transmission.

8.6 Quantum Computing and Quantum Cryptography

The operation of communications and electronic equipment in common use is based on the deterministic, bulk properties of materials. Recent research has been aimed at exploiting the statistical properties of quantum mechanics, developing systems for communications, and computing based on the behavior of single subatomic particles.

8.6.1 Quantum Computing

Quantum computing aims to make use of the basic quantum properties of the nucleus of atoms to make a powerful computer [16]. A conventional digital computer manipulates numbers expressed in binary form (i.e., in terms of bits, each of which takes the value 0 or 1). The spin of a nucleus can also take two values. Unlike the conventional computer, it can also take both values at once. This effectively allows a system to be stimulated with all of its possible input values simultaneously, permitting the parallel computation of all possible output values. It has recently been shown that such an approach can be used to speed up the factorization of numbers, greatly reducing the computational effort required for the cryptanalysis of many public-key encryption systems [17].

The development of practical quantum computers is still in its infancy. It is not yet known whether it is possible to build a system with sufficient precision to factorize large numbers. Even if it is feasible to build such systems, the bulk and cost of the support infrastructure may limit their use. The number of problems for which quantum computing is known to offer significant speed improvements over conventional computers is also small, although it includes a number of important areas such as searching and factorization of large numbers. The realization of low-cost quantum computers will in itself negate the value of most in-service security services based on public-key encryption algorithms.

8.6.2 Quantum Cryptography

Quantum cryptography, sometimes more accurately known as quantum key distribution (QKD) [18], provides a means for exchanging encryption keys with the unconditional security offered by a one-time pad. Once exchanged, these keys can be used with a secret-key encryption system or as a one-time pad.

Let us say that *A* wants to transfer a key to *B* (Figure 8.11). *A* randomly generates the key, and for each bit, sends a single photon with the value of

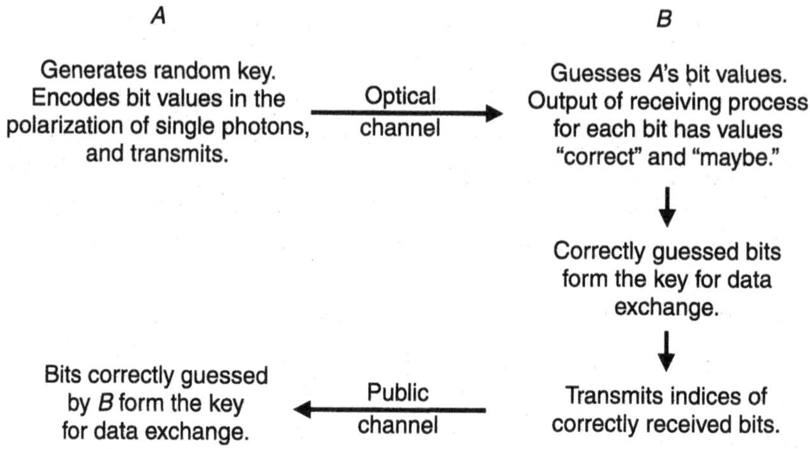

Figure 8.11 Key exchange process for quantum cryptography.

the bit represented in the polarization of this photon. A polarization of 90° may represent a 0, with 45° representing a 1. For each bit, *B* guesses the value that *A* has sent. If *B* guesses 0, *B* sets a polarizing filter to –45°; if *B* guesses 1, *B* sets the polarizing filter to 0°. If *B*'s filter is aligned with the polarization of *A*'s photon, it will pass and be detected. If *B*'s filter is perpendicular to the polarization of *A*'s photon, the photon is absorbed by the filter. Otherwise, there is a 50% chance that the photon passes through the filter and is detected. Therefore, if *B* correctly guesses *A*'s bit value, there is a 50% chance that the photon will be detected; if *B* incorrectly guesses *A*'s bit value, no photon will be detected. Hence, the detection of a photon implies that *B* has correctly guessed *A*'s bit value.

After *A* has transmitted a number of photons, *B* has detected some of these photons (on average, 25%), and therefore knows the correct value of the associated bits. *B* then tells *A* over a public communications channel which bits in the sequence *B* has received. These bits are then used as for encrypting data. However, *B* does not transmit the values of these bits over the public channel.

If an eavesdropper intercepts the photons between *A* and *B*, errors (with a rate of at least 25%) will be introduced into the transmission process. This situation will be detected quickly by *A* and *B* when they attempt to communicate using the secret key. They can then reinitiate the key exchange. They only way that an eavesdropper can subvert this process is to sit between *A* and *B* on both the channel carrying the photons and the public channel. This type of attack can be prevented by the use of an authenticated public channel, possibly using a small number of bits from a previous quantum key exchange for the authentication.

Errors on the transmission channel can be corrected using error-correcting codes, as discussed in Chapter 5. Because the introduction of errors is also associated with the presence of an eavesdropper, it is not possible to absolutely guarantee security when errors are corrected. It is possible, however, to calculate the probability that an eavesdropper has successfully intercepted the key transmission, and to use this probability to provide a statistical guarantee of security.

Key exchanges using this and similar protocols have been successfully carried out on optic fiber with ranges up to approximately 50 km. Optic fiber is an almost ideal transmission medium, preserving the polarization of photons and minimizing the number of interfering photons arriving at the receiver.

One of the key areas for future work is on the use of free-space quantum key distribution. Ranges up to 0.5 km have been reported with both transmitter and receiver at ground level in daylight with an error rate of 1.6%, and 0.2% of transmitted photons successfully detected [19]. The overwhelming majority of these errors (approximately 75%) were concluded to arise from imperfections in the optics used for the experiment. If successful, this work on free-space quantum cryptography could enable key exchange between a low-Earth-orbit satellite and a ground station.

8.6.3 Implications for EW

The major relevance of both quantum computing and quantum cryptography to EW is in the area of encryption. Quantum computing offers the promise of rendering useless at least those public-key encryption systems whose security is based on the factorization of large numbers. Quantum cryptography offers the potential of an alternative means for key exchange, offering unconditional security, at least on error-free channels.

One important unexplored area is the potential for jamming the transmission of a quantum key, which might be achieved by the jammer transmitting additional photons toward the receiver. If the transmission error rate can be increased sufficiently, the key sender (A) and receiver (B) will be unable to determine whether or not the key has been intercepted in the transmission path.

Endnotes

[1] Policy for the GIG is defined in:
U.S. Department of Defense Chief Information Officer Guidance and Policy Memorandum 10-8460, "GIG Network Operations," August 24, 2000.
U.S. Department of Defense Chief Information Officer Guidance and Policy Memorandum 7-8170, "GIG Information Management," August 24, 2000.
U.S. Department of Defense Chief Information Officer Guidance and Policy Memorandum 4-8460, "GIG Networks," August 24, 2000.

[2] "Specification of the Bluetooth System," Bluetooth SIG, Version 1.1, February 2001.

[3] As discussed for other commercial technologies in Chapter 5, the implementation of frequency hopping in Bluetooth is designed to provide protection against natural or unintentional interference. It is not design to protect against adversary jamming.

[4] Farber, D. J., "Predicting the Unpredictable: Technology and Society," in Anderson, R. H., *The Global Course of the Information Revolution: Technological Trends: Proceedings of an International Conference,* CF-157-NIC, Santa Monica, CA: Rand Corporation, 2000.

[5] Johnson, D. B., and D. A. Maltz, "Protocols for Adaptive Wireless and Mobile Networking," *IEEE Personal Communications,* Vol. 3, No. 1, February 1996.

[6] See, for example, Stallings, S., *Network and Internetwork Security,* 2nd Edition, Englewood Cliffs, NJ: Prentice Hall, 1995.

[7] This definition of confidentiality goes further than would usually be the case for a fixed network. It is required to encompass protection against traffic analysis.

[8] There is an unfortunate clash of terminology in the use of the term interception between ES and that used for network security.

[9] See, for example, Scholtz, R. A., "Multiple Access with Time-Hopping Impulse Modulation," in *Proc. MILCOM,* October 1993; or Win, M. Z., and R. A. Scholtz, "Impulse Radio: How It Works," *IEEE Communications Letters,* Vol. 2, No. 1, January 1998, pp. 10–12.

[10] Multiple Access Communications Ltd, "An Investigation into the Potential Impact of Ultra-Wideband Transmission Systems," U.K. Radiocommunications Agency, RA0699/TDOC/99/002, February 2000.

[11] Letter from Office of Spectrum Management, National Telecommunications and Information Administration to the Federal Communications Commission, imposing limits on the grant of a waiver to Part 15 of the FCC rules for ultra wideband radio, June 15, 1999.

[12] A waiver for limited use of UWB radio by fire and police departments in the United States was granted by the FCC on June 29, 1999. This waiver is subject to the use of UWB radio not interfering with other services.

[13] Standards defining HF ALE are:
FED-STD 1045A "Telecommunications: HF Radio Automatic Link Establishment," October 1993.
FED-STD 1046/1 "Telecommunications: HF Radio Automatic Networking Section 1, Basic Networking ALE Controller," October 1993.

FED-STD 1049/1 "Telecommunications: HF Radio Automatic Link Establishment in Stressed Environments, Section 1: Linking Protection," July 1993.

[14] Mitola, J., "Technical Challenges in the Globalization of Software Radio," *IEEE Communications Magazine,* Vol. 37, No. 2, February 1999, pp. 86–98.

[15] "Software Communications Architecture Specification MSRC-5000SCA," Joint Tactical Radio System (JTRS) Joint Program Office, Version 2.0, December 2000.

[16] Waldrop, M. M., "Quantum Computing," *Technology Review,* May 2000.

[17] Shor, P. W., "Polynomial-Time Algorithms for Prime Factorization and Discrete Logarithms on a Quantum Computer," *SIAM Journal of Computing,* Vol. 26, 1997, pp. 1484–1509.

[18] Singh, S., "Quantum Confidential," *New Scientist,* Vol. 164, October 2, 1999, pp. 28–33.

[19] See, for example, Hughes, R. J., et al., "Free-Space Quantum Key Distribution in Daylight," Los Alamos National Laboratory, Report No. LA-UR-99-737, 1999; and Hughes, R. J., et al., "Quantum Cryptography for Secure Communications to Low-Earth Orbit Satellites," Los Alamos National Laboratory, Report No. LA-UR-99-2741, 1999. Both of these reports also contain extensive bibliographies.

List of Acronyms

ALE	Automatic link establishment
AM	Amplitude modulation
ARQ	Automatic repeat request
ASK	Amplitude shift keying
C2	Command and control
C2W	Command and control warfare
C3	Command, control, and communications
CA	Certificate authority
CCD	Charge-coupled device
CDMA	Code-division multiple access
CIS	Communications and information systems
CNR	Combat net radio
CNRI	Combat net radio interface
COMSEC	Communications security
COTS	Commercial off-the-shelf
CP	Command post
CRC	Cyclic redundancy check
CSMA	Carrier-sense multiple access
CW	Carrier wave
DES	Data encryption standard
DEW	Directed energy weapon
DF	Direction finding
DSB	Double sideband
DSSS	Direct sequence spread spectrum

EA	Electronic attack
EOB	Electronic order of battle
ECCM	Electronic counter-countermeasures
ECM	Electronic countermeasures
EMCON	Emission control
EMI	Electromagnetic interference
EMP	Electromagnetic pulse
EMSEC	Emission security
EP	Electronic protection
EPLRS	Enhanced position locating and reporting system
ES	Electronic support
ESM	Electronic support measures
EW	Electronic warfare
EWLO	Electronic warfare liaison officer
FDMA	Frequency-division multiple access
FEC	Forward error correction
FFT	Fast Fourier transform
FH	Frequency hopping
FM	Frequency modulation
FSK	Frequency shift keying
GIG	Global information grid
GPS	Global positioning system
GSM	Global system for mobile communication
HELW	High-energy laser weapon
HF	High frequency
HPM	High-power microwave
HPRF	High power radio frequency
IDEA	International data encryption algorithm
IF	Intermediate frequency
IFM	Instantaneous frequency measurement
IO	Information operations
IP	Internet protocol
IPB	Intelligence preparation of the battlefield
IW	Information warfare
JTRS	Joint tactical radio system
KEA	Key exchange algorithm
LPD	Low probability of detection
LPI	Low probability of intercept

LSB	Lower sideband
MLRS	Multiple-launch rocket system
MRAAW	Medium-range anti-armor weapon
NATO	North Atlantic Treaty Organization
NBC	Nuclear, biological, and chemical
NCS	Net control station
NCW	Network-centric warfare
N-EMP	Nuclear electromagnetic pulse
NTDR	Near-term digital radio
PCS	Personal communications systems
PKA	Public key authority
PN	Pseudonoise
PSK	Phase shift keying
QAM	Quadrature amplitude modulation
QKD	Quantum key distribution
QPSK	Quadrature phase shift keying
RF	Radio frequency
RS	Reed-Solomon
RSA	Rivest-Shamir-Adleman
SCRA	Single-channel radio access
SEAD	Suppression of enemy air defense
SHF	Super-high frequency
SINCGARS	Single channel ground and air radio system
SOJ	Standoff jammer
SRAAW	Short-range anti-armor weapon
SSB	Single sideband
TCP	Transmission control protocol
TDMA	Time-division multiple access
UAV	Uninhabited aerial vehicle
UHF	Ultra high frequency
USB	Upper sideband
UWB	Ultra wide band
VHF	Very high frequency
VSB	Vestigial side band

About the Authors

Dr. Michael Frater is an associate professor in the School of Electrical Engineering at the University of New South Wales at the Australian Defence Force Academy. He has more than 10 years of experience in the development of communications systems and services, including videoconferencing and video and image surveillance. He has led a number of collaborative projects investigating image and video communications over low-bandwidth links. Dr. Frater has been actively involved in the development of international standards for audio-visual communications and broadcasting and has served as head of the Australian delegation to the Moving Picture Expert Group (MPEG), one of the major international standards bodies working in this area. He holds a B.Sc. and a B.E. in electrical engineering from the University of Sydney and a Ph.D. in systems engineering from the Australian National University. His research and teaching interests lie in digital audio-visual communications, including compression; transmission and delivery electronics; broadcasting and datacasting; telecommunications networks and architectures; and mobile communications (PCS technology and services). He is the author of a number of articles on communications systems and communications services.

Dr. Michael Ryan received his B.E., M.Eng.Sc., and Ph.D. in electrical engineering from the University of New South Wales, Canberra, Australia, in 1981, 1989, and 1996, respectively. Since 1981, he has held a number of positions in communications and systems engineering and in management and project management as a lieutenant colonel with the Royal Australian Signal Corps. Since 1998, he has been with the School of Electrical Engineering at the University of New South Wales at the Australian Defence

Force Academy, where he is currently a senior lecturer. His research and teaching interests are in communications systems (network architectures, electromagnetics, radio wave propagation, mobile communications, and satellite communications), information systems architectures, data compression for remote sensing applications, systems engineering, project management, and technology management. He is the editor-in-chief of the *Journal of Battlefield Technology*, and is the author of a number of articles on communications and information systems and a book on battlefield command systems.

Index

The Artech House Information Warfare Library

Electronic Intelligence: The Analysis of Radar Signals, Second Edition, Richard G. Wiley

Electronic Warfare for the Digitized Battlefield, Michael R. Frater and Michael Ryan

Electronic Warfare in the Information Age, D. Curtis Schleher

EW 101: A First Course in Electronic Warfare, David Adamy

Information Warfare Principles and Operations, Edward Waltz

Principles of Data Fusion Automation, Richard T. Antony

For further information on these and other Artech House titles, including previously considered out-of-print books now available through our In-Print-Forever® (IPF®) program, contact:

Artech House
685 Canton Street
Norwood, MA 02062
Phone: 781-769-9750
Fax: 781-769-6334
e-mail: artech@artechhouse.com

Artech House
46 Gillingham Street
London SW1V 1AH UK
Phone: +44 (0)20-7596-8750
Fax: +44 (0)20-7630-0166
e-mail: artech-uk@artechhouse.com

Find us on the World Wide Web at:
www.artechhouse.com

www.ingramcontent.com/pod-product-compliance
Lightning Source LLC
Chambersburg PA
CBHW071300180926
51613CB00001B/87
* 9 7 8 1 5 8 0 5 3 2 7 1 6 *